A Course in Ordinary Differential Equations with Applications

A Course in Ordinary Differential Equations with Applications

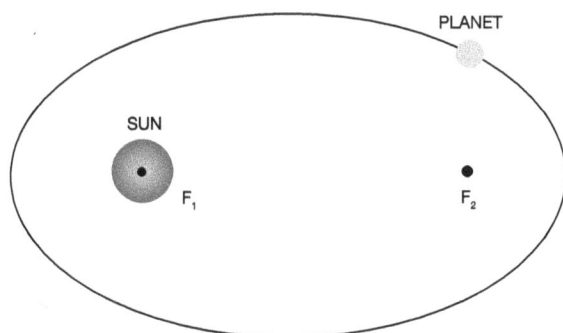

Martin A Moskowitz
Graduate Center of the City University of New York

World Scientific

NEW JERSEY · LONDON · SINGAPORE · GENEVA · BEIJING · SHANGHAI · TAIPEI · CHENNAI

Published by

World Scientific Publishing Co. Pte. Ltd.

5 Toh Tuck Link, Singapore 596224

USA office: 27 Warren Street, Suite 401-402, Hackensack, NJ 07601

UK office: 57 Shelton Street, Covent Garden, London WC2H 9HE

Library of Congress Control Number: 2024059687

British Library Cataloguing-in-Publication Data
A catalogue record for this book is available from the British Library.

A COURSE IN ORDINARY DIFFERENTIAL EQUATIONS WITH APPLICATIONS

ISBN 9789819801718 (hardcover)
ISBN 9789819801725 (ebook for institutions)
ISBN 9789819801732 (ebook for individuals)

For any available supplementary material, please visit
https://www.worldscientific.com/worldscibooks/10.1142/14066#t=suppl

Desk Editors: Soundararajan Raghuraman/Rok Ting Tan

Typeset by Stallion Press
Email: enquiries@stallionpress.com

Preface and Acknowledgments

This book was written for advanced undergraduate math or science majors. Its initial purpose was to illustrate the elementary mathematical theory of ordinary differential equations (ODEs) and their diverse and powerful applications. Historically, these have been decisive in many physical problems, some of which have philosophically challenged and indeed altered our civilization's concepts. As the reader will see, the names Newton, Leibnitz and Euler, who followed Copernicus, Galileo, and especially Kepler, have played a major role. Then came a number of inspired others, such as Bessel, Frobenius, Sturm, Liapunov, Liouville, Hermite, Hamilton, Picard, Poincare, and Volterra, who also each contributed heavily on the mathematical side of things. Here, we have limited ourselves primarily to differential equations of the first and second order as these, particularly second-order equations, are the heart of the matter, as well as being quite sufficient in terms of presenting the reader with challenges.

The first two chapters and the early part of the third are the core of the *historical* part of the subject and are chock-full of interesting and important applications to the "real world". They consist of mostly linear ODEs with constant coefficients of the first and second order, both homogeneous and non-homogeneous. Here, among many other things, we deal with planetary motion and prove Kepler's three laws as a consequence of Newton's law of gravitation together with his law of motion, $F = ma$. The latter part of Chapters 3 and 4 deal with equations with variable coefficients, gives an account of Hermite functions, the power series method of Frobenius, and is perforce somewhat technical. It could be omitted on a first reading.

v

Chapter 5 contains ODEs which are not solvable by quadrature, i.e., in terms of elementary functions, and so approximations need to be made. It contains the classical equation of the pendulum under *small* oscillations and the system of "predator-prey" equations of Volterra. There we also prove Liapunov's theorem and apply it to both the pendulum and predator–prey systems. Chapter 6 contains applications of ODE to differential geometry of surfaces, namely, Minding's theorem where the curvature is assumed constant, the Sturm comparison theorem where there are bounds on it, and as corollaries, we derive the geometric theorems of Bonnet and Hadamard. Although I have endeavored to cover all that is needed in that subject in order to make what we are doing completely intelligible to the reader (see also Chapter 12), for this chapter, it would be desirable to have had some contact with differential geometry beforehand.

We then discuss some more advanced topics. This is where we may not be able to explicitly solve an ODE, but can qualitatively describe the solutions. Chapters 7 and 8 deal with planar and general autonomous systems. This includes Lieonid's theorem which gives sufficient conditions for a class of second-order equations to have a periodic solution. Chapter 9 concerns the matrix exponential function, matrix differential equations and *systems* of first-order equations with both constant and variable but periodic coefficients. It also deals with the relationship of homogenous and non-homogeneous equations with variable coefficients. Although matrix differential equations is a subject which is properly addressed in a book on ODE, and one that I am particularly interested in because it is intimately connected to Lie theory, it too might be omitted on a first reading. However, in order to deal with this material, it would be necessary for the reader to be familiar with linear algebra, specifically the Jordan canonical form. Depending on where one's interest lay, the same could be said of Chapter 8 on the systems of linear ODEs and an introduction to autonomous dynamical systems culminating with the Poincare–Bendixson theorem, where solutions of ODEs in general cannot be found explicitly, but their mathematical qualities can be described. However, doing so is not recommended as this subject has many links with numerous applications in science, particularly involving *computer science* which a young reader would likely find interesting and so could be read with profit. Chapters 8 and 9 are linked in that the axioms for an autonomous dynamical system in Chapter 8 are theorems concerning the exponential function in Chapter 9. Then follows Chapter 10 on the classical partial differential equations (PDEs) of the second order in two variables and should be included in a first reading.

In Chapter 11, we turn to the calculus of variations. This subject matter is fundamental. Therefore, it is of importance to both advanced undergraduate students in mathematics or in one of the sciences, and as the reader will see, it is inextricably linked to ODE. The calculus of variations is a beautiful, powerful and venerable subject invented by Euler (and Lagrange, or perhaps by Bernoulli, or perhaps by Fermat.) But it was Euler who first understood its not so distant roots in calculus itself. However, here, rather than looking for an unknown point that minimizes or maximizes some given numerical function by finding its critical points, i.e., solving some numerical equation, in the calculus of variations, one extremizes some property of a function (which is after all a "point" in some function space) and so we will need to find its "critical points" which will mean solving some ODE (or system of ODEs or PDEs) in that function space. These ODEs are usually of the second order. Thus, such ordinary differential equations are built into the mathematics and often the physics of the calculus of variations. The last section of this chapter deals with problems in more than one independent variable and therefore PDEs culminating with the Plateau problem.

Our final chapter (Chapter 12) concerns curvature and the Gauss–Bonnet Theorem for two-dimensional surfaces which continues some of the issues of Chapter 6.

All these comments can be used to structure a course of study based on this book as a text depending on whether this is to be a leisurely year course for advanced undergraduates or only a semester which would depend on how ambitious the instructor is in terms of selecting material. Doing the entire book would be appropriate for a one-semester graduate course.

Finally, there are six appendices. These address the Gaussian distribution, Picard's fixed point theorem on the existence and uniqueness of local solutions for more or less all first-order ODEs, Stokes theorem in its general form, real analytic functions, Fourier series and lastly some remarks on special relativity. These appendices should make it easier for undergraduates to deal with the background material, which is considerable.

Let me say at the outset that this text will not bore the reader with numerous pointless exercises. Rather, it is noteworthy that a small number of types of differential equations have cut such a wide swath, there being few subjects in mathematics which can make such a claim. It is my privilege and pleasure to be able to introduce the reader to these and the many consequences that follow. It is for these reasons I have written it, with the idea of sharing my love of mathematics with my readers in the hope that they will have these same feelings about it. Although my friend and colleague, Dick Sacksteder, had nothing one could point to, to do with this endeavor, as he was truly an expert in this subject, I can't help but feel that (whether with approval or not) he has been looking over my shoulder as I worked on it.

This book and its author owe a considerable debt of gratitude to several people: The diagrams were kindly provided by Oleg Farmakis, the complex task of compiling the LaTeX file into what the reader now sees was done by Dr. Josiah Sugarman and the voluminous amount of proof reading by Andre Moskowitz. I wish to express my profound appreciation to each of them for their important contributions.

I want to thank my beloved wife, Anita, for her encouragement and forbearance during the course of the preparation of the manuscript. This book is dedicated to Stefan, Culver, Dillon and Hannah.

About the Author

Martin Moskowitz got his PhD in 1964 from the University of California, Berkeley, where his dissertation was written under the direction of Gerhard Hochschild. He held teaching positions at the University of Chicago and Columbia University before coming to the CUNY Graduate Center where he became Full Professor in 1975 and was Chair of the Mathematics Department there for six years. He successfully directed eight PhD students: two at Columbia and six at the Graduate Center.

On various occasions, he has visited the University of California, Berkeley, Paris VI, La Sapienza, Tor Vergata, the IHES, Darmstadt and SUNY Stonybrook. He was the recipient of a number of National Science Foundation grants, held a NATO Senior Post-Doctoral Fellowship in Science, and has been a Fellow at Centro Nazionale delle Ricerche, Italy.

His areas of specialization in mathematics are Lie groups and algebraic groups, particularly properties of the exponential map, representation theory and harmonic analysis and homogeneous spaces of finite volume. He was an Editor of the *Journal of Lie Theory* for over 20 years before stepping down in 2022 and has written over 50 research papers and seven books.

Contents

11. An Introduction to the Calculus of Variations

12. The Gauss Bonnet Theorem for Surfaces in $\mathbb{R}^3$

Chapter 1

An Introduction to First-Order Ordinary Differential Equations

In this chapter, we give a brief exposition of the early stages of ordinary differential equations (ODE) and the required attendant parts of differential and integral calculus. We then define an ODE and analyze what it means to solve one.

1.1 The Mean Value Theorem and Some of Its Consequences

We begin with the important mean value theorem (MVT) for derivatives (not to be confused with the MVT for integrals).

Theorem 1.1.1. *Let $f : [a, b] \to \mathbb{R}$ be a differentiable function. Then there exists an $\bar{x}$ (a mean value), where $a < \bar{x} < b$ such that*

$$\frac{f(b) - f(a)}{b - a} = \frac{df}{dx}(\bar{x}).$$

The special case where $f(a) = f(b)$ is called Rolle's theorem, which we shall prove first as it is the heart of the matter.

Proof of Rolle's Theorem. As we know, since f is differentiable, it is continuous and as $[a, b]$ is a closed and bounded interval, f achieves maximum $M = f(\alpha)$ and a minimum $m = f(\beta)$, where α and $\beta \in [a, b]$. If either α or β is an interior point, then α or β is a local min or max, and therefore, the derivative at that point is

1

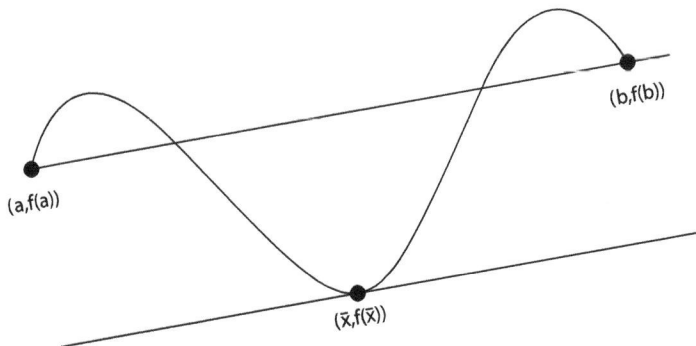

Figure 1.1.   The mean value theorem.

zero, proving Rolle's theorem in that case. The only other possibility is that both α and β occur at the end points. But if they occurred at the same end point, then $M = m$, and so, f would be constant. Hence, its derivative would be zero, so there again we would be done. The only remaining possibility is that each of them occur at different end points, say $M = f(a)$ and $m = f(b)$ (or vice versa). But since $f(a) = f(b)$, we would again get $M = m$, so again f is constant and its derivative is identically zero, completing the proof of Rolle's theorem. $\square$

Turning to the general case, consider the auxiliary function

$$g(x) = f(x) - \frac{f(b) - f(a)}{b - a} x$$

and observe that $g(a) = g(b)$, and since then, $g'(\bar{x}) = 0$ for some $\bar{x}$,

$$f'(\bar{x}) = \frac{f(b) - f(a)}{b - a}.$$

We remark what the MVT shows is if we draw the straight line joining $(a, f(a))$ and $(b, f(b))$ and take another line in the plane parallel to it, but far away (either above or below) and slowly move it toward the graph of f, while always keeping it parallel, when we first meet the graph, the line will actually be *tangent* to the graph!

We now turn to the question of the subject of this book: What is a differential equation? A *differential equation* is an equation involving an unknown function, say f, of a *single* (usually) *real* variable together with certain of its derivatives. Such an equation is called an *ordinary differential equation* or ODE, the term "ordinary" referring to the fact that we are dealing with functions of one variable. By the *order* of such an equation is meant the order of the highest derivative of f which appears in the equation. Here, we shall restrict ourselves to differential equations of the first or second order. This is not only because these are easier to solve but also because such ODEs come up in many interesting physical or other real-world problems (as well as in applications to other parts of mathematics), and so, there is a big payoff in studying them. As we shall see, when time t is the variable, they can predict the future.

A *solution* to an ODE is any function satisfying the equation; and solving the equation means finding *all* solutions. We shall leave the question of domains rather vague and content ourselves with the general remark that whatever the common domain of the functions given in the differential equation to be solved is, that should be the domain of a solution. That is, the solution to the differential equation should be valid throughout that domain. One final introductory remark: Since we are interested in all possible solutions to the given equation and solving such an equation ultimately involves integration, which perforce introduces an arbitrary constant of integration, as a rule of thumb we should expect that the number of arbitrary constants needed to describe the set of all solutions will be the same as the order of the equation. An initial condition is knowing (or insisting upon!) f or some derivative of some order taking a particular value at a point x_0 in the domain of f. It is very typical of this subject for an ODE of order n, after it is solved, that n initial data points will uniquely determine a solution. For this reason, we will want to know all the solutions of our equation. (However, as a word of caution to the reader, this is only possible in early stages of the subject. In general, it may not be possible to determine a solution in terms of known or easily described functions, but only to be able to give a qualitative description of it).

One further remark: At the cost of damping enthusiasm, it is easy to give examples of ODEs that simply have no solutions. For example, if $\mathrm{tr}(x, y)$ is the trace of a 2×2 positive, semi-definite, symmetric quadratic form in x and y, then because the eigenvalues,

λ and $\mu \geq 0$, we see

$$\operatorname{tr}(x, y) = \lambda x^2 + \mu y^2 \geq 0.$$

Evidently, the ODE

$$\frac{dy}{dx}^2 + \operatorname{tr}(x, y) + 1 = 0$$

can have no (real) solutions. In particular, none of the following first-order ODEs have solutions. $\frac{dy}{dx}^2 + x^2 + y^2 + 1 = 0$, $\frac{dy}{dx}^2 + x^2 + 1 = 0$, $\frac{dy}{dx}^2 + y^2 + 1 = 0$.

As we know, the derivative of a constant function is zero. The following shows the converse is also true. Thus, we will have solved our first differential equation, namely, $f' = 0$.

Corollary 1.1.2. *Let f be a differentiable function on $[a, b]$. If $\frac{df}{dx}$ is identically zero on (a, b), then f is constant on $[a, b]$.*

Proof.   By the MVT, there exists $\bar{x}$ with $a < \bar{x} < b$ satisfying the equation above. But since $\frac{df}{dx}(\bar{x})$ must be 0, $f(a) = f(b)$. Now, take α to be a place on $[a, b]$, where f takes a maximum value M and β where f takes a minimum value m. Thus, $f(\alpha) = M \geq f(\beta) = m$. But reasoning as above, we get $f(\alpha) = f(\beta)$. Hence, $M = m$, and so, f is constant on (a, b). But since f is continuous on $[a, b]$, the conclusion follows. $\square$

Of course, this corollary shows if f and g are smooth functions defined on the same interval, $[a, b]$ and if $\frac{df}{dx} = \frac{dg}{dx}$ and f and g agree at some point, then $f = g$ identically on $[a, b]$. This is because $(f - g)' = f' - g' = 0$.

Exercise 1.1.3. Use induction to show that if the nth derivative, $f^n = 0$ for some positive integer n, then f is a polynomial of degree $\leq n$. Indeed, if n is the smallest such integer, then $\deg f = n$. In particular, if the derivative is constant, then the function is linear and if the derivative is linear, i.e., if the second derivative is constant, then the function is a quadratic polynomial, etc.

So, for example, if a body is falling because of the force of gravity, then since $F = ma$, we see that its weight, $W = mg = m\frac{dv}{dt}$, where $v(t)$ is its velocity at time t. Then the ms cancel and $\frac{dv}{dt} = g$, the surface gravity. Since its derivative is constant, $v(t)$ is linear $v(t) = gt + v_0$, where v_0 is the initial velocity (say at $t = 0$). Therefore, the position $s(t)$ is quadratic $s(t) = \frac{1}{2}gt^2 + v_0 t + c$. Choosing coordinates so that $c = s(0) = 0$, finally we get $s(t) = \frac{1}{2}gt^2 + v_0 t$. This is the mathematics of Galileo's experiment with falling bodies at the Campanile of Pisa Cathedral.[1] Note the trajectory is independent of the weight, W. Also note that the formula is valid whether the object is thrown upward or downward. For example, if it is thrown upward, v_0 takes a negative sign.

We next turn to a rather militeristic application by applying these ideas to the question of the range of a cannon at ground level with muzzle velocity v_0. If θ is the angle the cannon makes with the ground, then $v_0 \sin(\theta)$ is the vertical component of the initial velocity v_0. Let $Y(t)$ be the vertical height of the cannon ball at time t. Since $F = ma$, taking account of the fact that the force is opposite to gravity and canceling m, we see $-g = \frac{d^2Y}{dt^2}$. Hence, $\frac{dY}{dt} = -gt + c$, where c is a constant. Thus, $\frac{dY}{dt} = -gt + v_0 \sin(\theta)$, and so, because we are at ground level, $Y(t) = -\frac{1}{2}gt^2 + v_0 \sin(\theta)t$ (a parabola). The range occurs when $Y(t) = 0$ and $t > 0$, that is, at time $t = \frac{2v_0 \sin(\theta)}{g}$.

To find the range, we need to know $X(t)$, the horizontal component of motion at time t. Since there are no forces acting horizontally $\frac{dX}{dt} = v_0 \cos(\theta)$ and since the initial position horizontally is also zero, $X(t) = v_0 \cos(\theta)t$. Hence, the range R is given by

$$R = \frac{2v_0 \sin(\theta)}{g} \cdot v_0 \cos(\theta).$$

Since $\sin(2\theta) = 2\sin(\theta)\cos(\theta)$, we see

$$R = \frac{v_0^2}{g} \sin(2\theta),$$

that is, the range is independent of the mass of the projectile and depends only on the muzzle velocity and, of course, θ. Moreover, for

---

[1]Galileo (1564–1642) was the noted Italian Renaissance physicist who empirically studied falling bodies, the pendulum (see Chapter 5), and turned the telescope toward the heavens.

any cannon, the maximal range occurs when $\theta = \frac{\pi}{4}$ and its maximal range is $\frac{v_0^2}{g}$.

As we shall now see, even these trivial ODEs have important applications.

1.2   Conservation of Energy and the Escape Velocity

In this section, we consider the situation of the force F of gravitational attraction between two bodies which is given by

$$F = -\frac{GmM}{r^2},$$

where m and M are the respective masses of the two bodies, r is the distance between their centers of gravity and G is the universal gravitational constant (valid throughout the universe). The notion of the existence of this force is due to Issac Newton. This is a vector equation, and the vector is pointing from m to M. Therefore, if we take coordinates with M as the origin, there needs to be a minus sign.

More generally, we could consider any force F as being a vector field defined and smooth on the domain consisting of all of $\mathbb{R}^3$, *excluding* the origin. (As we shall see in a moment, under gravitational attraction, everything actually takes place on a line.) We call a force field F *conservative* if its value depends only on the position, the equation of gravitational attraction being clearly such a field. Let E denote the total energy, namely, the sum of the potential energy and the kinetic energy of a system. As we shall see shortly, if F is conservative, then E must be constant in time along the entire trajectory $\gamma(t)$ of motion. This is called *the principle of conservation of energy*. We mention that in Appendix F, where we discuss special relativity, when objects travel at or near the speed of light, this principle is violated in a rather spectacular way by equation F.

Whether F is conservative or not, the potential energy is

$$f(b) = -\int_0^b F(u)du,$$

where the integral is along the path $\gamma(t)$ joining the points $(0, \gamma(0))$ and $(b, \gamma(b))$ in question. On the other hand, the *kinetic energy* of a particle of mass m depends on things other than the position. It is $\frac{1}{2}m|v|^2$, where v is the velocity and $|v|$ the speed. (The point of this distinction is that the velocity is a vector, while the speed is its scalar length.) If a particle of mass m travels along a trajectory $\gamma(t)$, the vector velocity at time t is $\gamma'(t)$ and the acceleration is $\gamma''(t)$. Newton's law of motion is then expressed by the vector equation,

$$F(\gamma(t)) = m\gamma''(t).$$

Thus, $E(t)$, the total energy at time t, is

$$E(t) = f(\gamma(t)) + \frac{1}{2}m \left\| \gamma'(t) \right\|^2.$$

The *principle of conservation of energy* is the following.

Theorem 1.2.1 (Principle of Conservation of Energy). *For any smooth curve $\{\gamma(t) : t \in [a,b]\}$, the total energy of a particle traveling on such a path under the effect of a conservative field of forces is constant.*

Proof. Let us calculate $\frac{dE}{dt}$ at the point $\gamma(t)$. Using the *chain rule* and the formula for differentiating the scalar product of two vector functions,

$$\frac{d}{dt}\langle Y(t), Z(t) \rangle = \langle Y'(t), Z(t) \rangle + \langle Y(t), Z'(t) \rangle.$$

Hence,

$$\frac{dE}{dt} = \langle f(\gamma(t)), \gamma'(t) \rangle + m\langle \gamma'(t), \gamma''(t) \rangle,$$

that is, $\frac{dE}{dt} = \langle -F(\gamma(t)) + m\gamma''(t), \gamma'(t) \rangle$. Taking account of Newton's law, this is $\langle 0, \gamma'(t) \rangle = 0$. Thus, by Corollary 1.1.2, E is constant along the curve.   □

Now, suppose we are interested in the physical system consisting of a single point particle of mass m located on the Earth's surface given an initial velocity and subject only to the force of the earth's gravitational field. Will it get pulled back to the surface? In particular, this would apply to a rocket given an initial impetus with no other forces exerted on it other than gravity. We ask how fast must it be going initially in order to not be pulled back to earth by gravity. This speed is called the *escape velocity*. To answer this question, we shall apply the principle of conservation of energy. We remark that at first glance, it is not obvious that there is any speed for which this would occur.

We may assume that the object leaves the earth on some radial line, so here we have a one-dimensional problem. Also, from the above, when the rocket is a distance r from the center of the earth, the gravitational force on it is given by $F(r) = \frac{k}{r^2}$. In order to evaluate the positive constant k, observe that when $r = R$, the radius of the earth, that force is its weight, mg. Thus, $k = mgR^2$. Let $v(r)$ be the velocity at position r. The principle of conservation of energy tells us that the constant total energy is given by

$$E = \frac{1}{2}mv^2(r) - \int_R^r F(u)du.$$

We can determine this constant by taking the variable r to be R. This yields $E = \frac{1}{2}mv^2(R) = \frac{1}{2}mv_0^2$. Thus, $\frac{1}{2}mv_0^2 = \frac{1}{2}mv^2(r) - \int_R^r F(u)du$. Evaluating the integral and taking into account the value of k, we see that

$$\int_R^r F(u)du = mgR^2 \left(\frac{1}{R} - \frac{1}{r} \right).$$

Substituting this into the above equation and dividing by m yields

$$v_0^2 - v^2(r) = 2gR^2 \left(\frac{1}{R} - \frac{1}{r} \right).$$

The fact that m doesn't appear in this equation shows that, whatever be the outcome, it must be *independent of m*. Now, the object falls back if and only if at some point $v(r) = 0$. Alternatively, to escape, we need to only be sure that $v(r) > 0$, for all r, as $r \to \infty$. But, in that case,

$$v_0^2 > v_0^2 - v^2(r) = 2gR^2 \left(\frac{1}{R} - \frac{1}{r} \right),$$

and as $r \to \infty$, the right side of this equation approaches $2gR$. Thus,

$$v_0 \geq \sqrt{2gR}.$$

Conversely, if this is so, or even if $v_0 = \sqrt{2gR}$, then $v_0^2 = 2gR$, so

$$v_0^2 - v^2(r) = 2gR - v^2(r) = 2gR^2 \left(\frac{1}{R} - \frac{1}{r} \right).$$

Thus, $v^2(r) = \frac{2gR^2}{r}$, and so, $v(r) > 0$. Therefore, the proof is complete.

Corollary 1.2.2. *The earth's escape velocity is $\sqrt{2gR}$.*

Some remarks: A direct calculation using the above formula yields approximately 3.5 mi/sec. The reader should also note that the units here are right, namely, $\sqrt{ft/\sec^2 \cdot ft} = ft/\sec$. Of course, if we were on a different body such as the Moon, then this formula would still be valid, it would only be necessary to take g to be the Moon's surface gravity and R to be the Moon's radius.

1.2.1   The distance to the horizon and a property of the circle

Some of these exercises do not involve ODEs, but in spite of this, in the view of the author, they seem to fit in rather well with the material. Here, we shall make heavy use of the following theorem which concerns an important property of a circle. We remark that this could also easily be proved by Calculus and the fact that two lines are perpendicular if and only if their slopes are negative reciprocals of one another. Our proof uses the notion of *transitivity*, which is important and useful. The reader will note that in this way, the theorem actually proves itself!

Theorem 1.2.3. *The tangent line to a point, p, on a circle is perpendicular to the radial line to p.*

Proof. There is no loss of generality in assuming that the circle is centered at the origin since any circle can be translated so that its center is the origin and translations preserve tangency and perpendicularity. Let $x^2 + y^2 = r^2$ be its equation. Since the circle is radially symmetric and rotations preserve tangency and perpendicularity, there is also no loss of generality in choosing any point on the circle since any point on the circle can be gotten from any other point by a rotation and as we said rotations preserve tangency and perpendicularity. Therefore, we can take $p = (0, r)$ the point where our circle meets the positive y-axis. The equation of the tangent line to the circle at this point is $y = r$ and the radial line to this point is the x-axis which is perpendicular to it.   □

This is an example of the important concept that here the isometry group acts *transitively*. One sees that translations act transitively on the plane and rotations act transitively on the circle. Note that for the argument to work not only does one need transitivity, but also these transformations must preserve whatever the property is that one wants to prove.

Another way to go in proving the theorem above would be the following. The reader should do it as an exercise, keeping in mind that here if and only if means if the intersection of a line with the circle consists of a single point, then the line is tangent to the circle at that point and conversely if the line is tangent to the circle at some point p, then the intersection of this tangent line and the circle consists only of p (and nothing else).

We leave as an exercise for the reader the fact that the intersection of a line l and a circle consists of either 0, 1 or 2 points. l is tangent to the circle if and only if the intersection consists of a single point.

We now turn to questions about distance to the horizon due to the *curvature* of the Earth.

Exercise 1.2.4. Suppose you are on a dock at the edge of the sea that is h feet above sea level (say at high tide). Assuming the Earth is a perfect sphere of radius R (and you are lying on the dock rather than standing), how far away is the horizon?

Solution: Let t be the distance to the horizon, s be the intercepted arc along the Earth and u be the straight line distance from the foot of the dock to the horizon (see diagram.) Then $(R+h)^2 = R^2 + t^2$ and since h, s, t forms a right triangle and so for small h, $t^2 = h^2 + s^2$, and so, $(R + h)^2 = R^2 + h^2 + s^2$. Hence, for small h compared to R,

$$s = \sqrt{2Rh}.$$

Alternatively, since for small h as compared to R, t is approximately s. Since $(R + h)^2 = R^2 + t^2$, $2hR + h^2 = t^2$ which is approximately s^2. Moreover, $2hR + h^2$ is approximately $2Rh$ when h is small compared to R, so we again get $s = \sqrt{2Rh}$ approximately.

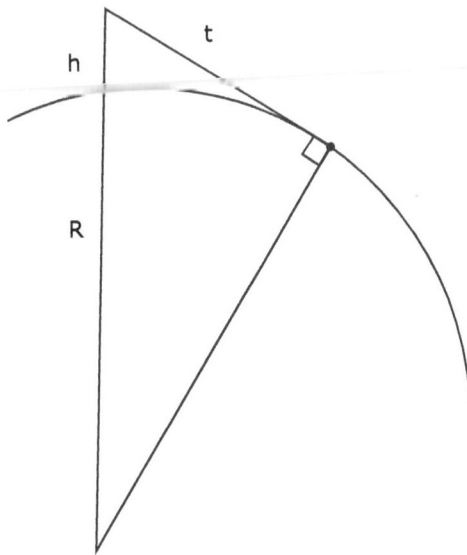

Figure 1.2. The distance to the horizon when viewed from a dock.

In our next application, h is not so small as compared to R. Here, we ask what is the minimal height a rocket fired vertically has to go to see all of the continental US?

Exercise 1.2.5. Assume the radius R of the earth is approximately 4000 miles, and so, the circumference is approximately 25000 miles. Taking the width of the continental US (which is larger than the height) as 2900 miles, if a rocket is fired vertically upward from the geographic middle of the US (say somewhere in Kansas), how far up does it have to go in order to just see all of the US?

Solution: Use the diagram of the previous problem and let ϕ be the central angle at the center of the earth made by the lines from the foot of the dock and the foot at the horizon to the center of the Earth is given by $\frac{\phi}{360} = \frac{s}{R}$, where $s = 1450$ mi. Use this to calculate ϕ. Then calculate t by observing $\tan(\phi) = t/R$. Now, since $(R + h)^2 = R^2 + t^2$. Hence, $h^2 + 2Rh - t^2 = 0$. Solving this quadratic equation for the height (and rejecting the negative root), $h = R(\sec(\theta) - 1)$. Then the angle is $\theta = \frac{2900 \cdot 360}{2 \cdot 25000} = 21$ degrees. Therefore, h is approximately 280 miles.

1.3   Some Further Developments: The Exponential Function

We shall now apply these ideas to solve a less trivial, but important ODE, namely, $f' = \lambda f$ (on the entire real line, $\mathbb{R}$). To do so, we calculate the derivative of $\frac{f(x)}{Ce^{\lambda x}}$, where $C \neq 0$ and λ are constants. A direct calculation using the quotient rule tells us it is

$$\frac{e^{\lambda x} f' - f \lambda e^{\lambda x}}{Ce^{2\lambda x}}.$$

(Of course, here we are assuming $\lambda \neq 0$, since when $\lambda = 0$, we would be back in the case of the previous ODE).

From the ODE, we see that the numerator is identically zero. Therefore, as we now know, the function $\frac{f(x)}{Ce^{\lambda x}}$ is constant. Hence,

$$f(x) = Ce^{\lambda x}.$$

Since the converse is easily checked by differentiation to also hold, we have solved this ODE. It gives us an alternative characterization of the exponential function (as the solution to this ODE). The use of this ODE is not just philosophical decoration; it has many applications. Here are some of them: When $\lambda > 0$, the equation keeps track of *bacterial growth* since for bacteria in a petri dish at constant temperature and with available food, the rate of growth, λ (which depends on the type of bacteria and environmental features), is proportional to the number of bacteria. Similarly, instantaneous compound interest satisfies this differential equation. On the other hand, if one has *radioactive decay*, the rate of growth is proportional (with a negative constant of proportionality) to the amount of radioactive material present. Here, λ varies with the radioactive element. In this way, one can solve such problems if we know λ and an initial condition (say when $x = 0$).

Another application of the ODE we now consider involves *barometric pressure*. Atmospheric pressure, p (measured by a barometer say in lbs./sq. in.) at a given point on the Earth's surface, or above it, varies with the height h above the surface. The Gas law, which involves the relationship of absolute (Kelvin) temperature, T, and p, creates difficulties, so we make the bold assumption that T is independent of p, i.e., independent of h (which is very dubious!). However, it gets us the following ODE, where R is the radius of the Earth:

$$\frac{dp}{dh} = -k \left(\frac{R}{R+h} \right)^2 p(h).$$

Now, in most applications, R, which is about 4000 miles, is much larger than h, so the factor $\left(\frac{R}{R+h} \right)^2$ is very slightly less than 1. Taking it to be 1,[2] we finally arrive at $\frac{dp}{dh} = -kp(h)$, so that $p = Ae^{-kh}$.

An alternative way to look at our method of solution of these rather simple ODEs is to consider separation of variables when that is possible. Thus, since $\frac{df}{dx} = \lambda f(x)$, we can write $\frac{df}{f} = \lambda dx$, by integrating both sides and setting these equal, we get, for some constant D,

$$\log(f(x)) = \int \left(\frac{df}{f} \right) = \int \lambda dx = \lambda x + D.$$

This means that $f(x) = e^{\lambda x + D} = e^D e^{\lambda x}$ and taking $C = e^D$ proves what we need.

More generally, these are the examples of the method of separation of variables, which works as follows: If the ODE is anything of the form $\frac{dy}{dx} = g(x)h(y)$, where the functions g and h are smooth functions of x and y, respectively, and h is never zero, then separating the variables $\frac{dy}{h(y)} = g(x)dx$ and integrating, we see that $\int_{\text{Dom }Y} \left(\frac{dy}{h(y)} \right) = \int_{\text{Dom }X} g(x)dx + \text{Constant}$. If we can solve this last equation for y, we would have solved the ODE.

Example: As an interesting application of these ideas, we ask if the reader has ever noticed that when a faucet is turned on and water descends into the sink, the stream of water has a smaller and smaller diameter. Let us analyze why.

First, we name the cast of characters. Let t be the time that the stream of water leaves the faucet, whose cross-section we assume is circular. We consider a cylinder of this stream of small height. Let $A(t) = \pi r(t)^2$ be its area, where $r(t)$ is its radius, and $y(t)$ its vertical distance from the opening of the faucet. Then $V(t)$, its

---

[2]Note the similarity here to the problem above concerning the distance to the horizon.

volume, is $A(t)y(t)$. Since the mass (or weight W) of our cylinder must be conserved over time and $\delta = W/V$, which is also constant, this means $V(t)$ itself is constant as t varies. Thus, $\frac{dV}{dt} = 0$. By the formula for the derivative of a product,

$$A(t)y'(t) + A'(t)y(t) = \pi r(t)^2 y'(t) + 2\pi r(t)y'(t) = 0,$$

and since $r(t)$ is everywhere positive,

$$\frac{r'(t)}{r(t)} = -\frac{1}{2}\frac{y'(t)}{y(t)}.$$

Thus, $r'(t)$ is negative everywhere, confirming our observation that $r(t)$ is always decreasing. But we really want a quantitative statement of how fast $r(t)$ is decreasing! This will result from solving this ODE which is in parametric form, i.e., both r and y are functions of t. (We could just as well have regarded r as a function of y to put the ODE in the more familiar form, but this will give the reader some experience in dealing with ODEs in parametric form.) Integrating 1.3, we see $\log(r(t)) = -\frac{1}{2}\log(y(t)) + c$. Suppose, for example, when $y = 1$ cm, $r = r_1$. Then since $\log(1) = 0$, $c = \log(r_1)$ and $r(t) = r_1 y(t)^{-1/2}$. Thus, when water falls from a faucet, the radius decreases as follows:

$$r = \frac{r_1}{\sqrt{y}}.$$

In a similar spirit, the following is a useful tool for constructing global diffeomorphisms. A homeomorphism $f : X \to Y$ is a bijective and bicontinuous function. Here, we shall assume X and Y each have a metric, making it more convenient to talk about continuity. This would be the case if, for example, X and Y were subsets of $\mathbb{R}^n$. Similarly, a diffeomorphism $f : X \to Y$ is a bijective map where f and f^{-1} are continuously differentiable functions. That is, all derivatives up to some order exist and are continuous. The example of X and Y being subsets of $\mathbb{R}^n$ is key here as well.

Theorem 1.3.1. *Let $f : \mathbb{R} \to \mathbb{R}$ be a differentiable function such that for all x, $f'(x) \geq \epsilon$ where $\epsilon > 0$. Then f is a global diffeomorphism.*

In particular, if $f(x) = \lambda \cdot x + \phi(x)$, where $\lambda > 0$ and $\phi : \mathbb{R} \to \mathbb{R}$ is a differentiable function with $\phi'(x) > 0$, then since $f'(x) = \lambda + \phi'(x)$, $f' \geq \lambda > 0$ everywhere and so is a global diffeomorphism.

Proof.   Since $f' > 0$, f is a local diffeomorphism. In particular, f is an open map and $f(\mathbb{R})$ is an open set in $\mathbb{R}$. By continuity, $f(\mathbb{R})$ is also connected. This means $f(\mathbb{R})$ is either a point, an open interval, an open half-line or $\mathbb{R}$ itself. Moreover, since $f'(x) > 0$, f is strictly increasing and so is injective. In particular, $f(\mathbb{R})$ can't be a point. If $f(\mathbb{R})$ were either an open interval, or an open half-line, then since f is strictly increasing, it would have to have a horizontal asymptote. By the mean value theorem, this would mean $f'(x) \leq \frac{\epsilon}{2}$ at various points, a contradiction. Hence, f is also surjective. □

Exercise 1.3.2.

- The reader should check that $f(x) = \tan^{-1}(x)$ and $f(x) = \tanh(x)$ are examples, showing that it is not enough in Theorem 1.3.1 to merely assume $f'(x) > 0$ for all $x \in \mathbb{R}$.
- Check that $\phi(x) = \sum_{i=1}^{n} \lambda_{2i+1} x^{2i+1}$, where $\lambda_{2i+1} \geq 0$ has $\phi'(x) \geq 0$.
- So, for example, let $\phi(x) = x^{2n+1}$, where n is a positive integer. Note, however, that this function is not a diffeomorphism of $\mathbb{R}$ (Why not?) but is a homeomorphism (Why?).
- Use the intermediate value theorem to show that any polynomial of odd degree maps surjectively (but perhaps not injectively) onto $\mathbb{R}$. Hence, if p is a polynomial of odd degree and $p'(x) > 0$ everywhere, then p is a global diffeomorphism of $\mathbb{R}$.
- Find all the times when the minute hand and the hour hand of a clock coincide.

As a final example of *separation of variables*, we consider the ODE, $\frac{dy}{dx} = 1 + y^2$, over some real interval on the x-axis. Since $1 + y^2$ is always positive and so never zero, $\frac{1}{1+y^2}$ is a smooth function. Separating the variables gives us $\frac{dy}{1+y^2} = dx$. Integrating, we see $\tan^{-1}(y) = x - c$ and so $y(x) = \tan(x - c)$. However, if $x - c = \pm\frac{\pi}{2}$, this solution is not defined. Indeed, it takes the value $\pm\infty$ and so taking an interval centered at c with end points, $-\frac{\pi}{2}$ and $\frac{\pi}{2}$, we see that there is no interval of length $\geq \pi$ where a solution exists, but within a well-chosen interval of length π, it does.

1.4   Further Developments Involving First-Order Equations

Beyond the exponential function itself, another very important function, fundamental in probability and statistics, called the *normal Gaussian*[3] *distribution* is given by

$$f(x) = e^{-x^2}.$$

This function arises in many contexts. Its graph is the "bell-shaped curve" of probability and statistics. (More generally, one can consider

$$g(x) = Ce^{\lambda x^2},$$

where C, λ are constants.) The chain rule tells us that f satisfies the ODE $f'(x) = -2xf(x)$.

Exercise 1.4.1. Use separation of variables as above to show that the solution to this ODE is $f(x) = Ce^{-x^2}$ for some suitable positive constant C (which we can take to be $f(0)$).

We now consider a slightly more general (non-homogeneous) first-order ODE, namely,

$$\frac{df}{dx} = \lambda f + \mu,$$

where the constant $\lambda \neq 0$ and μ is another non zero-constant. Note when $\mu = 0$, this is the (homogeneous) equation we just solved. As the reader should check, the method used to solve the homogeneous equation doesn't work anymore. If we want to solve this ODE, we shall need another idea. This is the use of an *integrating factor* due to Leibnitz (1646–1716). He and Newton (1643–1727) were the inventors of Calculus.

---

[3]Gauss, Carl Friedrich (1777–1855) was one of the world's great mathematicians with numerous pioneering contributions to many fields in mathematics, physics and astronomy.

Since the purpose of the integrating factor is to find something which is an *exact derivative*, we now give a characterization of this concept called the *Poincaré Lemma*.[4]

Proposition 1.4.2. *Let D be a simply connected region in the plane, $\mathbb{R}^2$ and $F(x, y) = (p(x, y), q(x, y))$ be a C^1 vector-valued function on D. Then $F = \mathrm{grad}(g)$ for some smooth numerical function g on D if and only if*

$$\frac{\partial p}{\partial y} = \frac{\partial q}{\partial x}$$

everywhere on D.

Simple connectivity is a topological concept which intuitively means in this space any simple smooth closed curve can be continuously deformed (*always within the space*) to a point.

If D were not simply connected, for example, if the planar region D a hole, or even a single point missing, then Proposition 1.4.2 would not imply $F = \mathrm{grad}\, g$. A typical example of this is given on p. 425 of [10]. Actually, Proposition 1.4.2 holds in all dimensions, see, e.g., [10]. Its proof depends on line integrals. When it exists, the function g is called a *conservative potential*. Of course, going one way is trivial and requires no special hypothesis on D, for if $\frac{\partial g}{\partial x} = p$ and $\frac{\partial g}{\partial y} = q$, then $\frac{\partial^2 g}{\partial y \partial x} = p_y$ and $\frac{\partial^2 g}{\partial x \partial y} = q_x$. Since g is assumed to be smooth, taking partial derivatives with respect to x and y commute and so the above equation holds.[5]

After we have solved equation (1.4), we will apply the solution to the question of a falling body, but this time with a parachute. This will make precise why such a person doesn't just crash and die! Note that if one is foolish enough to jump out of a plane without a parachute, say from rest, or more likely if the parachute fails to open, it follows from $F = ma$ that the ODE is $f''(t) = g$. Therefore, his speed, $f' = gt + b$, increases without bound and the result would not be pretty.

---

[4]Henri Poincare (1854–1912) was a student of Hermite and the leading French mathematician of his generation. He is usually credited with inventing topology. The Institute HP in Paris is named after him.

[5]In topological terms, what this says is that if D is simply connected, its first homotopy group is trivial, while the first homotopy group of $D \backslash \{p\}$ is non-trivial.

However, rather than merely solving the equation just above, it will pay to go for a little more and solve the following more general ODE:

$$\frac{df}{dx} = p(x)f + q(x),$$

where $p(x)$ and $q(x)$ are given smooth *functions*. To do this, for any smooth function, $h(x)$, we define $\int h(x)dx$ to be any function whose derivative is $h(x)$. In general, the existence of such a thing is guaranteed by the fundamental theorem of calculus. Actually, in simple cases, it is not necessary to appeal to general theorems; in fact, it is *not even desirable since we are interested in explicit solutions*. For example, if $p(x) = \lambda$, a constant, then $\int p(x)dx = \lambda x$ (rather than $\lambda x + C$).

Now, suppose f is a solution to the equation above. Multiply f pointwise by $e^{-\int p(x)dx}$ (this latter term is called an integrating factor) and differentiate using the product rule.

$$\frac{d}{dx}(fe^{-\int p(x)dx}) = \frac{df}{dx}e^{-\int p(x)dx} + fe^{-\int p(x)dx}(-p(x)).$$

Since f is a solution, we see that

$$\frac{d}{dx}(fe^{-\int p(x)dx}) = q(x)e^{-\int p(x)dx}.$$

Integrating, we get

$$fe^{-\int p(x)dx} = \int q(x)e^{-\int p(x)dx}dx + C,$$

where C is a constant. Finally, multiplying both sides of this last equation by $e^{\int p(x)dx}$ and solving for f yields

$$f(x) = e^{\int p(x)dx}\int q(x)e^{-\int p(x)dx}dx + Ce^{\int p(x)dx}.$$

Thus, we have shown that all solutions to the original equation are of the form above. Clearly, all steps of our argument are *reversible*; hence, this proves that the set of all solutions to the given equation consists precisely of those of the form above. Observe that there is one arbitrary constant involved in the set of solutions.

What this amounts to in the specific case we are interested in is the following: Suppose $p(x) = \lambda$ and $q(x) = \mu$, both non-zero constants. That is, we are solving the differential equation,

$$\frac{df}{dx} = \lambda f + \mu,$$

where λ and μ are constants. After performing the requisite integration in the general solution, taking into account that $\lambda \neq 0$, we get as the set of all solutions

$$f(x) = -\frac{\mu}{\lambda} + Ce^{\lambda x}.$$

We now apply the solution of this equation to the situation of retarded fall, such as what happens to a parachutist where there is a retarding force proportional to the velocity. Here, instead of x, we take, for the independent variable, time, t. The fundamental equation of motion $F = ma$ in this case yields $mg - kv = m\frac{dv}{dt}$, where m is the mass of the falling body and $mg = W$ is its weight (of the parachutist), $v(t)$ is its vertical velocity at time t, $\frac{dv}{dt}$ is the acceleration and k is a positive constant depending on the size of the parachute and perhaps its material. (We consider the parachute itself to be weightless.) Then dividing by m, we get

$$\frac{dv}{dt} = -\frac{k}{m}v + g.$$

Thus, here $\mu = g$ and λ is the negative number $-\frac{k}{m}$. As we now know, the general solution is $v(t) = \frac{W}{k} + Ce^{-\frac{k}{m}t}$. For example, if the parachutist leaves the plane from rest, i.e., with vertical velocity zero, then we can use this "initial condition" to determine C and thus get the *unique* solution to our problem; since $v(0) = 0$, $C = -\frac{W}{k}$ and the general solution is

$$v(t) = \frac{W}{k}(1 - e^{-\frac{k}{m}t}).$$

In particular, $v(t) < \frac{W}{k}$. Since the solution holds for all time $t \geq 0$, letting $t \to \infty$ tells us that as $t \to \infty$,

$$v(t) \to \frac{W}{k}.$$

This number is called the *terminal velocity*. Its existence is the reason that these people don't crash and die. Note that neglecting the weight of the parachute itself, as we may, a person of double the weight will have a terminal velocity *double* that of the lighter person.

Another situation which is subject to the same type of ODE has to do with artillery, or in more modern terms, rocketry. Consider a rocket whose trajectory is the path $(x(t), y(t))$ in a vertical plane launched at the origin, $(0,0)$, at an angle θ, with the horizontal and initial velocity v_0. Assume that air resistance is always acting contrary to the trajectory and proportional to the speed with proportionality constant $k > 0$. Then $v(t) = (x'(t), y'(t))$ and $v(0) = (x'(0), x'(0)\tan(\theta))$. Since $F = ma$ in each coordinate, we get

$$m\frac{d^2x}{dt^2} + k\frac{dx}{dt} = 0$$

and

$$m\frac{d^2y}{dt^2} + k\frac{dy}{dt} - mg = 0.$$

Each of these can be solved *independently*. To do so, just let $z = \frac{dx}{dt}$ or $\frac{dy}{dt}$ to get first-order equations of the type we have been considering and apply the initial conditions. Then put them together to get the trajectory of the rocket. These details are left to the reader as an exercise.

The ODEs under considerations also work very well when applied to problems connected with temperature or air pressure change. Newton's *law of cooling* states that if a body is heated to some high temperature and then placed in an environment of lower temperature, u_0, then as it cools over time t, $u(t)$ satisfies

$$\frac{du}{dt} = -k[u(t) - u_0].$$

Here, k is some positive constant depending on the properties of the material being heated. Then as we know, $u(t) = u_0 + Ae^{-kt}$, where A is a constant which depends on the initial temperature of the object.

Question: If one is making coffee and wants it to remain nice and hot when served sometime later, is it better to add cold milk immediately or exactly when it is served?

As a further application, we have the Traitrix. This is a curve in the x, y plane gotten by dragging a point A on the positive x-axis at the end of a string, AT, of length, say a, as T moves upward along the y-axis.

Evidently, the ODE of the Traitrix is

$$\frac{dy}{dx} = \frac{-\sqrt{a^2 - x^2}}{x}.$$

Separating the variables tells us

$$y = -\int \frac{\sqrt{a^2 - x^2}}{x} dx.$$

Since when $x = a$, $y = 0$, as an exercise, we ask the reader to verify

$$y = a \log \left[\frac{a + \sqrt{a^2 - x^2}}{x} \right] - \sqrt{a^2 - x^2}.$$

Looking to the future, as a final application to first-order equations, we solve the differential equation

$$\frac{df}{dx} = \lambda f + A e^{\mu x},$$

where λ, μ and A are constants. Here, $p(x) = \lambda$ and $q(x) = A e^{\mu x}$. As we know, the general solution to this ODE is $f(x) = A e^{\lambda x} \int e^{(\mu - \lambda)x} dx + C e^{\lambda x}$. However, it will now be necessary to distinguish two cases, namely, $\lambda = \mu$ and $\lambda \neq \mu$.

As can be checked directly, if $\lambda = \mu$, the general solution is

$$f(x) = e^{\lambda x}(Ax + C),$$

where C is a constant, whereas if $\lambda \neq \mu$, the general solution is

$$f(x) = \frac{A}{\mu - \lambda} e^{\mu x} + C e^{\lambda x}.$$

Chapter 2

Planetary Motion

In this chapter, we turn to what is historically one of the world's most important (and controversial) scientific discoveries: that of how the planets move about in space. The heliocentric theory was first proposed by Copernicus in the 16th century and Galileo in the early 17th century and were not well received[1,2] The subject was then empirically and exhaustively studied by Kepler[3] later in the 17th century who formulated his *Three Laws of Planetary Motion*. These were then proved by Newton to be a consequence of his Universal Law of Gravitation, together with his General Law of Motion, $F = ma$. It is this last development that we deal with in what follows.

As we shall see, our analysis applies not only to planets orbiting the sun but also to a moon orbiting a planet as well as comets passing through the solar system, perhaps never to return, and it will differentiate between those comets that have periodic (elliptical) orbits and those that just pass through and never come back which,

---

[1]Copernicus was afraid of being accused of heresy (for which the penalty was, after torture to extract a confession, to be burnt at the stake) and so hesitated to publish for many years, while Galileo was merely placed under house arrest for the remainder of his life (this only because he had a personal relationship with the Pope).

[2]The reader may wonder why this was considered such a heresy. One reason for this was the fact that the Pope thought he must be at the center of the universe and not on some minor planet flittering about hither and yon.

[3]Johannes Kepler (1571–1630) was a German astronomer and natural philosopher, and was a key figure in the 17th century Scientific Revolution.

as we shall see, have hyperbolic orbits. Indeed, the Inverse Square Law is ubiquitous in physics as it not only appears in mechanics, but also in electricity and magnetism, as well as the study of atomic particles; our analysis actually applies to all of these situations.

In proving that these two physical laws of Newton imply those of Kepler, we shall (have to) make a number of simplifying assumptions. These include the assumption that the planet and the sun are each point masses. That is, for each, the total mass is concentrated at a point (their respective centers of gravity). Moreover, we shall assume that no other object (such as the moon or other planets) exerts any gravitational force on the planet (even though this is not true). This is known as the *two-body problem*. Hence, the force of gravitational attraction between the two is given by Newton's Universal Law of Gravitation:

$$F(r) = -\frac{GmM}{r^2},$$

where G is a universal (gravitational) constant (i.e., doesn't merely apply to our solar system), M the (larger) mass of the sun, m the (smaller) mass of the planet and r the distance between these point masses. This equation gives the *magnitude* of the force vector F on the earth and depends only on the distance, r, between the two. The *direction* of the vector F is along the line joining the two point masses, pointing toward the sun, which we shall take as the origin of coordinates. Since it's an attracting force, this vector takes a minus sign. We shall further assume that M is much larger than m. This has the effect of making the planet do all the moving rather than both moving about the *common center of gravity of the assemblage* consisting of the two bodies *which is actually what happens*. We describe the position in space of the planet as time t varies by the vector-valued function $X(t)$. Then $X'(t)$ is its velocity at time t. In particular, as explained just above, $F(t) = c(t)X(t)$, where $c(t)$ is some smooth numerical function of t. On the other hand, the law of motion tells us $F = ma(t) = mX''(t)$. Hence, $X''(t) = \frac{c(t)}{m}X(t)$ and so as t varies $X''(t)$ is itself a variable, but also a smooth multiple of $X(t)$. Thus, we have what seems to be a second-order homogeneous equation, but with *variable* coefficients.

We first verify that the planets actually move in a fixed plane rather than flitting about in space. The reader will note that the following lemma doesn't use the inverse square aspect of the Universal Law of Gravitation, but merely the fact that the force always points in the radial direction. However, as we shall see later, the inverse square aspect will actually determine *what kind of planar orbit one gets*.

Lemma 2.0.1. *If $X''(t)$ is a (variable) multiple of $X(t)$ for all t, then as t varies $X(t)$ lies in a fixed plane.*

Proof. Consider the vector-valued function $t \mapsto X(t) \times X'(t)$ (cross-product) and calculate its derivative,

$$\frac{d}{dt}(X(t) \times X'(t)) = X'(t) \times X'(t) + X(t) \times X''(t).$$

Each of these terms is zero because in both cases the angle between the vectors is zero. It follows from our result above (applied to each coordinate) that $X(t) \times X'(t)$ itself is the constant vector, and in particular, its direction is constant. But since $X(t) \times X'(t)$ must also be perpendicular to the plane determined by $X(t)$ and $X'(t)$ (at time t), that plane must have constant unit normal, i.e., be a fixed plane. In particular, $X(t)$ (and $X'(t)$) lies in it as t varies. □

We now coordinatize this plane (with the sun at the origin) by letting i and j be the unit vectors in the positive x- and y-directions, respectively. It will also be convenient to introduce polar coordinates (whose units are u_r, the radial, and u_θ, the tangential unit vector) and take the x- and y-axis in such a way that when $t = 0$ the planet is closest to the sun and lies on the positive x-axis. (Life experience tells us that there is such a minimum positive distance, i.e., the planets do not fly into the sun.)
Thus,

$$X(t) = r(t)\cos(\theta(t))i + r(t)\sin(\theta(t))j,$$

and $X(0) = r(0)i$, where there are no smaller values of $r(t) = |X(t)|$ than $r(0)$. Due to the *minimality* of $r(0)$, it follows by ordinary calculus which we ask the reader to verify that $X'(0)$ is perpendicular to $X(0)$ and so $X'(0) = cj$.

As we have just seen, $X(t) \times X'(t)$ is a constant vector, say C. We now show that this fact implies Kepler's first law, namely, equal areas are swept out in equal times (Kepler's first law also being independent of the Inverse Square Law of Gravitation). The area formula for polar coordinates shows the area $A(t)$ swept out between t and t_1 is given by

$$A(t) = \int_{\theta(t)}^{\theta(t_1)} \frac{1}{2} r^2(\theta) d(\theta).$$

By the fundamental theorem of calculus and the chain rule,

$$A'(t) = \frac{1}{2} r^2(\theta(t)) \frac{d(\theta(t))}{dt}. \tag{2.1}$$

Resolving the various vectors, i.e., F, v, a into their radial and tangential components and calculating $A''(t)$ by differentiating this, we get

$$A''(t) = \frac{1}{2} r^2(\theta(t)) \frac{d^2(\theta(t))}{dt^2} + r \frac{dr}{dt} \frac{d(\theta(t))}{dt}$$

$$= \frac{r}{2} \left[r(\theta(t)) \frac{d^2(\theta(t))}{dt^2} + 2r \frac{dr}{dt} \frac{d(\theta(t))}{dt} \right],$$

the term in the second bracket being the *normal component* of the *acceleration vector*. Since the force, and therefore the acceleration vector, is directed toward the origin, the normal component is zero, so $A''(t) = 0$, and therefore, as we observed earlier, $A'(t)$ is a constant, proving Kepler's first law as follows.

Corollary 2.0.2. $X(t)$ *lies in a plane and sweeps out equal areas in equal times.*

It also shows us that the acceleration equals its radial component and is

$$\frac{r}{2} \left[r(\theta(t)) \frac{d^2(\theta(t))}{dt^2} + 2r \frac{dr}{dt} \frac{d(\theta(t))}{dt} \right].$$

Up to now, we have not used the inverse square aspect of the law of gravitation, but merely that the force points in the radial direction. We shall now use the full strength of the law of gravitation, but first, we slightly rewrite it as

$$m \frac{d^2 X}{dt^2} = -\frac{GmM}{|X|^2}.$$

Canceling m, calling the *positive* constant $GM = k$ and writing $\frac{dX}{dt}$ as $v(t)$, we get

$$\frac{dv}{dt} = -\frac{k}{r^2}u_r. \tag{2.2}$$

Our task is now to solve what appears to be a (vector-valued) first-order differential equation with variable coefficients from which we hope to derive Kepler's second law, namely, the planets move in elliptical orbits with the sun at one of the foci. The fact that m cancels tells us, whatever the result is going to be, it will be independent of the mass of the planet, *but will depend on the mass M of the sun* and so will vary from one solar system to another.

Going back to the formula for $A'(t)$, we have

$$r^2(t)\frac{d(\theta(t))}{dt} = h, \tag{2.3}$$

where h is a constant. Since $r^2(t)\frac{d(\theta(t))}{dt} = dA(t)$, the infinitesimal area swept out at time t, we know it's constant, and because this is the rate at which area is swept out, we know $h > 0$. Since $dA = \frac{1}{2}hdt$, integrating, we see that $A(t_1) - A(t_2) = h(t_1 - t_2)$.

As we saw,

$$\frac{d^2r}{dt^2} - r\left(\frac{d\theta}{dt}\right)^2 = -\frac{k}{r^2}, \tag{2.4}$$

where $k = GM$. This nasty looking second-order ODE can be made to look more familiar if we change variables. Let $z = \frac{1}{r}$ and consider θ to be the independent variable rather than t. To see what new form this ODE takes, we first calculate $\frac{dr}{dt}$. We remind the reader that $r(t) = |X(t)|$.

$$\frac{dr}{dt} = \frac{d}{dt}\left(\frac{1}{z}\right) = \frac{-1}{z^2}\frac{dz}{d\theta}\frac{d\theta}{dt} = \frac{-1}{z^2}\frac{dz}{d\theta}\frac{h}{r^2} = -1h\frac{dz}{d\theta},$$

and so

$$\frac{d^2r}{dt^2} = -h\frac{d}{dt}\frac{dz}{d\theta} = -h\frac{d}{d\theta}\left(\frac{dz}{d\theta}\right)\frac{d\theta}{dt} = -h\frac{d^2z}{d\theta^2}\frac{-h}{r^2} = -h^2z^2\frac{d^2z}{d\theta^2}.$$

When this second derivative is inserted, we get

$$-h^2 z^2 \frac{d^2 z}{d\theta^2} - \frac{-1}{z} h^2 z^4 = -kz^2.$$

That is to say,

$$\frac{d^2 z}{d\theta^2} + z = \frac{k}{h^2}.$$

Thus, by good luck, we now have a *non-homogeneous second-order equation with constant coefficients*; in fact, since the coefficient of the linear term is positive, the homogeneous part is the harmonic oscillator equation whose solution is well known and is given in the following chapter. Indeed, by the lemma above concerning non-homogeneous equations, the solution to this equation is

$$z = A \sin(\theta) + B \cos(\theta) + \frac{k}{h^2}.$$

Now, continue with polar coordinates, but first shifting by rotating the x-axis so that it coincides with the line connecting (the point masses) m with M. Thus, the line $\theta = 0$ contains the point where the minimum distance between them is attained. From this, it follows that at that point $\frac{dz}{d\theta} = 0$ and $\frac{d^2 z}{d\theta^2} > 0$. But $\frac{dz}{d\theta} = A \cos(\theta) - B \sin(\theta)$, and when $\theta = 0$, this is A, so $A = 0$ and $\frac{dz}{d\theta} = -B \sin(\theta)$. Hence, $\frac{d^2 z}{d\theta^2} = -B \cos(\theta)$. Since this must be negative at $\theta = 0$, $B > 0$. Now, we cunningly re-replace z by $\frac{1}{r}$ and get $\frac{1}{r} = B \cos(\theta) + \frac{k}{h^2}$ so that

$$r = \frac{h^2/k}{1 + e \cos(\theta)},$$

where e is the positive constant, Bh^2/k.

   This is the equation of a (non-degenerate) conic with eccentricity e with a focus at the origin. Thus, the sun is at one of the foci. Now, if $e < 1$, this is an ellipse. If e is exactly 1, it's a parabola, and if $e > 1$, it's a hyperbola. Of course, e can never be exactly 1, especially since our many assumptions would make this very blurry indeed! Since by observation the planets are all periodic, in this case, we always get an ellipse. This is Kepler's second law as given in the following.

Corollary 2.0.3. *The planets move in elliptical orbits (in the plane of motion) with the sun at one of the foci.*

Most of the remainder of this chapter is more or less taken from
[13]. Since here we have an ellipse, an immediate consequence is that
$0 < e < 1$. However, we must keep in mind that when some comets
enter the solar system with high enough velocity (i.e., kinetic energy)
to escape, then $e > 1$, and therefore, these must follow *hyperbolic*
trajectories and hence will never come back! We will now prove this
statement. To do so, we need to find a *physical* quantity that determines the nature of the orbit. That quantity is the total energy, E,
of the system. This means that we must understand the relationship
of the mathematical quantity $e = \frac{Bh^2}{k}$ and the physical quantity, E,
which will require us to navigate a slew of constants.

As observed above, $\frac{1}{2}mv^2$, the kinetic energy of the planet, is

$$\frac{m}{2}\left[r^2\left(\frac{d\theta}{dt}\right)^2 + \left(\frac{d^2r}{dt^2}\right)\right],$$

while the potential energy of the system is the negative of the work
needed to move m to infinity where the potential energy is zero.
Thus,

$$-\int_r^\infty \frac{km}{r^2}dr = \frac{km}{r}\Big|_r^\infty = -\frac{km}{r}.$$

Therefore, E, the total energy of the system, is

$$\frac{m}{2}\left[r^2\left(\frac{d\theta}{dt}\right)^2 + \left(\frac{d^2r}{dt^2}\right)\right] - \frac{km}{r}.$$

Now, at the instant $\theta = 0$, $r = \frac{h^2/k}{1+e}$ and $E = \frac{mr^2}{2}\frac{h^2}{r^4} - \frac{km}{r}$. Due to the
principle of conservation of energy, we can calculate the total energy
at this instant. As the reader can check, substituting r from the first
equation into the second and solving the result for e yields

$$e = \sqrt{1 + E\left(\frac{2h^2}{mk^2}\right)}.$$

This enables us to write the equation of the orbit in terms of E (and
the other constants) rather than e. The equation of the conic then

becomes

$$r = \frac{h^2/k}{1 + E(2h^2/mk^2)\cos(\theta)}.$$

Since all the other constants involved are *positive*, the trajectory is an ellipse or a hyperbola according to whether $E < 0$ or $E > 0$. That is, the type of orbit (but not the orbit itself) is completely determined by the total energy E which we know to be constant by the principle of conservation of energy. The planets all have $E < 0$, while some of the comets have $E > 0$. They are the ones that don't come back. Finally, note that the actual orbit does depend on m, so it varies from one planet to another.

We now turn to *Kepler's third law*. Taking as given that we are now in the elliptical case, we investigate the period, ω, i.e., the planet's year.

Let a be one half the major axis and b one half the minor axis of our ellipse. After translation to the origin, in rectangular coordinates, the equation of the ellipse is

$$\frac{x^2}{a^2} + \frac{y^2}{b^2} = 1.$$

Exercise 2.0.4. Check that the total area of the ellipse is πab.

Classical analytic geometry tells us that $e = \sqrt{a^2 - b^2}/a$ and so

$$b^2 = a^2(1 - e^2).$$

On the other hand, in polar coordinates, the equation of the ellipse is

$$r = \frac{h^2/k}{1 + e\cos(\theta)}.$$

This means that

$$a = \frac{1}{2}\left[\frac{h^2/k}{1+e} + \frac{h^2/k}{1-e}\right] = \frac{h^2/k}{1-e^2} = \frac{b^2 h^2}{ka^2},$$

and so $b^2 = \frac{h^2 a}{k}$.

Now, the total area is of the ellipse is πab. On the other hand, as we observed, $A'(t)$ is constant $h\omega/2$. So, dividing the area by the period ω gives this constant rate of change of area, so $\pi ab = h\omega/2$, and therefore,

$$\omega^2 = 4\pi^2 a^2 b^2/h^2 = 4\pi^2 a^3/k.$$

Thus, ω^2 is proportional to a^3, the constant of proportionality being $\frac{4\pi^2}{k}$.

This law is true in any solar system. However, because of the presence of M, the constant of proportionality varies from one solar system to another since $k = GM$. But in any given solar system, all the planets have the same constant of proportionality. This is Kepler's third law. A further remark: Since $k = GM$ and $G = 6.6743 \cdot 10^{-11}$ m^3/kgsec^2, one could, for example, use our Earth's data and Kepler's third law to calculate the mass, M, of the sun!!

We conclude this chapter with the following exercises and observation.

Exercise 2.0.5. *How large does the moon appear?*

An interesting aspect of the movement of the moon about the earth is the question of where (or when) the distance between the two bodies is minimal so that the moon appears at its largest. Here, we consider the points, p, on an ellipse (with two distinct foci) and the question of when p assumes the shortest distance, $d(p)$, from a fixed choice f_0 of one of the foci. Of course, for this to happen, such a p must be a critical point of the function $d(p)$. We observe that here this means the line joining p and f_0 is *perpendicular* to the tangent line to the ellipse at p (prove). Now, there are exactly four possible points p on the ellipse which could fulfill this requirement. These are the two end points of the major axes p_1 and p_2. and two points symmetrically placed with respect to the minor axes p_3 and p_4 (see diagram in Figure 2.1).

Exercise 2.0.6. Prove that the point p_1 at the far end of the major axis from f_0 gives a global maximum distance while the point, p_2, at the near end of the major axis gives a local maximum, while the other two critical points p_3 and p_4 each give a *global minimum*. Thus, there are exactly two points on the ellipse where a minimal distance from f_0 is achieved. Figure 2.1 shows a rough sketch of the graph of the (periodic) function, $d(p)$.

What we have done in this chapter is referred to in the locker rooms of math and physics as the two-body problem (earth and sun). But there is also a secondary use of that term by these same people. That is the difficulty of finding an academic job in the same city for each of the spouses of a married couple and if they succeed one says

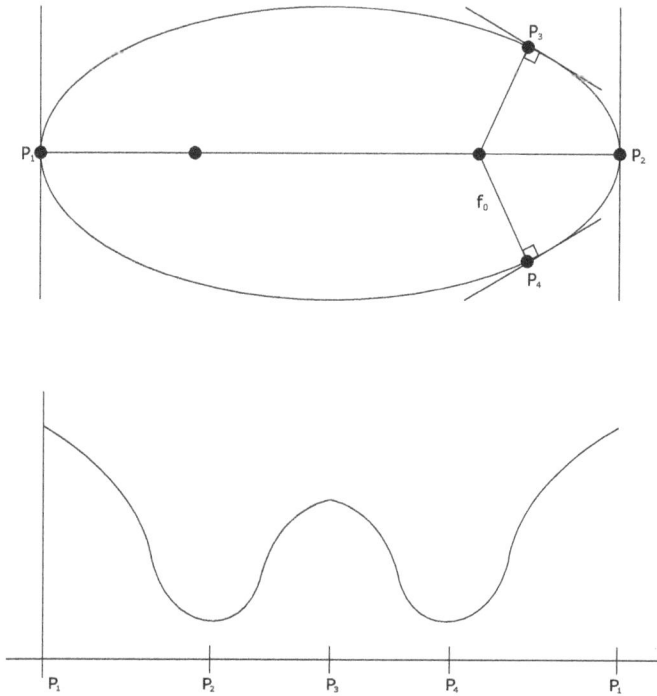

Figure 2.1.   How to determine when the moon is closest to the earth.

"they have solved the two-body problem". There is also "the three-body problem" where there are two planets and a sun (or the different situation of one planet, sun and moon). The three-body problem is what attracted Henri Poincare to ODE. The complications here are formidable and have never been resolved!

Chapter 3

Second-Order Ordinary Differential Equations

In this chapter, we first turn to some rather simple, but quite important, second-order equations with constant coefficients. We shall see, in one case, that because we were able to explicitly solve the rather general first-order equations, whose coefficients are *functions*, we will be able to parlay this into the solution of some second-order equations with *constant coefficients*. An example of this is the equation

$$\frac{d^2 f}{dx^2} - \omega^2 f = 0,$$

where ω is a constant. Of course, here, $\omega \neq 0$, since as we have already remarked, the trivial case where $\omega = 0$ has for solutions all linear functions and only those. Now, suppose f is a solution to this equation. Then it is easy to see that

$$\left(\frac{d}{dx} - \omega I \right) \left(\frac{d}{dx} + \omega I \right) f = 0.$$

Here, we *compose* these (commuting) operations. Such operations are called *differential operators*. Letting $g = (\frac{d}{dx} + \omega I)f$, we see that the equation merely says $(\frac{d}{dx} - \omega I)g = 0$, i.e., $\frac{dg}{dx} = \omega g$ and, as we observed earlier, this means that $g(x) = Be^{\omega x}$, where B is some constant. But then $\frac{df}{dx} = -\omega f + (-B)e^{\omega x}$. This is exactly the equation we have solved earlier, where $\lambda = -\omega$, $\mu = \omega$ and $A = -B$. Since $\omega \neq 0$, we see $\lambda \neq \mu$ and the general solution is $f(x) = \frac{-B}{2\omega}e^{\omega x} + Ce^{-\omega x}$. Since B and C are arbitrary constants, we can, by renaming them, see that the general solution to $\frac{d^2 f}{dx^2} - \omega^2 f = 0$ is

$$f(x) = ae^{\omega x} + be^{-\omega x}.$$

Conversely, the reader can check by differentiation that any such function is a solution. Note that these solutions are *unbounded* and are *not periodic*.

Exercise 3.0.1. Show that the general solution to this equation can also be expressed as

$$f(x) = A\sinh(\omega x) + B\cosh(\omega x),$$

where A and B are arbitrary constants.

Exercise 3.0.2. Try using the techniques established so far to solve the more general second-order equation with constant coefficients $a \neq 0$, b and c,

$$a\frac{d^2 f}{dx^2} + b\frac{df}{dx} + cf = 0,$$

under the assumption that the *associated quadratic equation* $aX^2 + bX + c = 0$ has only real roots. (This situation is a generalization of the one we have just dealt with.) Here, there are two cases to consider: the associated quadratic equation has either equal or unequal roots. Then show, by differentiation, that conversely any such function you have obtained is a solution. Then show that the solutions in the exercise just above form a real vector space of dimension 2.

We now turn to a similar looking second-order homogeneous ODE with constant coefficients, namely,

$$\frac{d^2 f}{dx^2} + \omega^2 f = 0,$$

where ω is a non-zero constant. Here, because of the plus sign, we shall see that the solutions are quite different from the situation of a minus sign. Indeed, here, the solutions turn out to be *periodic and bounded.* Note that here we have two complex conjugate (in fact pure imaginary) roots. This equation is called the harmonic oscillator equation.

To solve it, we first consider the simplest case of the equation, namely, when $\omega = 1$. If f is a solution to this equation, and we let $g = f'$, then $g' = -f$. But then, the derivative of

$$f^2 + g^2 \text{ is } 2ff' + 2gg' = 2fg - 2gf = 0.$$

As we know, since $f^2 + g^2$ has an identically zero derivative, it must be a constant; $f^2 + g^2 = C$. Clearly, unless f and g are both identically zero, $C > 0$, so $C = \alpha^2$ for some positive real number α.

Thus, $f' = \sqrt{\alpha^2 - f^2}$, where $\alpha > 0$. But then, integrating, we get $x = \arcsin(\frac{f}{\alpha}) - C$, where C is some (other) constant. Applying the sin function tells us $f(x) = \alpha \sin(x + C)$. Finally, the addition formula for sin shows that $f(x) = \alpha(\sin x \cos C + \cos x \sin C)$. Taking $A = \alpha \cos C$ and $B = \alpha \sin C$, we see that $f(x) = A \sin x + B \cos x$.

Now, suppose ω is arbitrary, but non-zero. Let $g(x)$ be any solution to this equation and $f(x) = g(\frac{1}{\omega}x)$. A direct computation shows that

$$\frac{d^2 f}{dx^2} = \frac{1}{\omega^2}\frac{d^2 g}{dx^2}\left(\frac{1}{\omega}x\right).$$

Hence,

$$\frac{d^2 f}{dx^2}(x) + f(x) = \frac{1}{\omega^2}\frac{d^2 g}{dx^2}\left(\frac{1}{\omega}x\right) + g\left(\frac{1}{\omega}x\right)$$

$$= \frac{1}{\omega^2}\left(-\omega^2 g\left(\frac{1}{\omega}x\right)\right) + g\left(\frac{1}{\omega}x\right) = 0.$$

By the first case we dealt with, this means that there exist constants A and B such that for all real x, $f(x) = g(\frac{1}{\omega}x) = A \sin x + B \cos x$. Letting $u = \frac{1}{\omega}x$, i.e., $x = \omega u$, where u is arbitrary, we see that $g(u) = A \sin(\omega u) + B \cos(\omega u)$. Since we are now down to questions of notation, any solution must be of the form $g(x) = A \sin(\omega x) + B \cos(\omega x)$. Conversely, by differentiating, it follows immediately that any such function is a solution.

Exercise 3.0.3. What is the amplitude and the period of the function

$$f(x) = A\sin(\omega x) + B\cos(\omega x)?$$

The equation we have just solved is called the *simple harmonic oscillator equation* for the following reason. Consider an object, which we assume to be concentrated at a point, hanging from a spring. We also assume that the force exerted by the spring is much larger than the weight of the object, and therefore, the weight can be ignored by comparison. (Alternatively, just let the weight of the object stretch out the spring to whatever it does and start from there). This simplifying assumption partially accounts for the term *simple*. The other reason for this term is that there is no *damping* which would contribute a first-order term to the ODE similar to what occurs in the parachute problem earlier. Let $x(t)$ be the vertical position, i.e., the distance at time t of its center of gravity from the equilibrium point. Now, Hooke's law tells us that the force the spring exerts on the point mass is proportional to the displacement and opposite to the direction of motion. If $k > 0$ denotes the constant of proportionality, then the fundamental equation of motion $F = ma$ tells us that after dividing m,

$$\frac{d^2x}{dt^2} + \omega^2 x = 0,$$

where $\omega = \sqrt{\frac{k}{m}}$. Thus, the general solution is $x(t) = A\sin(\omega t) + B\cos(\omega t)$, where A and B are constants. To interpret the meaning of these constants, let $t = 0$. Then $x(0) = B$. Having determined B, if we knew, say, $x(t_0)$ at some other time t_0, or if we knew the initial velocity $v(0) = \frac{dx}{dt}(0)$ (or at some other point), we could easily also determine A. In the latter case, $A = \frac{v(0)}{\omega}$. This is another example of a specific solution being determined by initial conditions. Suppose one does not wish to make the simplifying assumption of negligible weight. If $W = mg$ is not negligible compared to the force exerted by the spring, then the equation of motion is $mg - kx = m\frac{d^2x}{dt^2}$. In this case, we have a so-called *inhomogeneous* equation,

$$\frac{d^2x}{dt^2} + \omega^2 x = g,$$

where again $\omega = \sqrt{\frac{k}{m}}$. Now, by inspection, one sees that the constant function $x(t) = \frac{g}{\omega^2} = \frac{W}{k}$, where W is the weight of the object, is a solution to the inhomogeneous equation. Since the solutions to the homogeneous equation form a vector space (this also works when there is a $+$),

$$x(t) = A\sin(\omega t) + B\cos(\omega t) + \frac{W}{k}$$

is also a solution. As we observed, $x(t) = A\sin(\omega t) + B\cos(\omega t)$, where A and B are arbitrary constants, is the *most general* solution to the homogeneous equation, $\frac{d^2 x}{dt^2} + \omega^2 x = 0$. Therefore, we indeed have the most general solution to the inhomogeneous equation. (Of course, if $W = 0$, there would be no distinction between the solutions of the homogeneous and inhomogeneous equations.)

More generally, we have just proved the following.

Lemma 3.0.4. *The general solution of the inhomogeneous equation, $\frac{d^2 x}{dt^2} + \omega^2 x = c$, where ω and c are non-zero constants, is*

$$x(t) = A\sin(\omega t) + B\cos(\omega t) + \frac{c}{\omega^2}.$$

Before ending this section, we will mention, without proof, a general fact concerning the relationship between homogeneous and the related non-homogeneous ODEs. (But here we are referring to ODEs of any degree and with possibly variable coefficients.) Also, here, we consider a single equation although the result holds even for systems. For the details of this, see Section 2 of Chapter 9. However, before doing so, we need to make a remark about the roots of polynomials in one variable with real coefficients. First, note that the roots need not be real. For example, the polynomial $p(x) = x^2 + 1$ has all real coefficients (indeed, they are positive integers), but has no real roots since $x^2 + 1$ can never be 0 because when x is real, $x^2 + 1 \geq 1$. Indeed, the roots of this polynomial are $\pm i$. However, thanks to the *fundamental theorem of algebra*, which states that any polynomial of positive degree over $\mathbb{C}$ factors completely into linear factors. It is for this reason that when dealing with differential operators, we shall have to work over $\mathbb{C}$. In Chapter 7, we will prove the fundamental theorem of algebra (Corollary 7.1.2).

Again, assuming the fundamental theorem of algebra, as an exercise, we ask the reader to prove that in a polynomial in one variable of degree ≥ 1 with real coefficients, any non-real roots occur in complex conjugate pairs.

Here is the unproven assertion just mentioned. A proof can be found in Section 6 of Chapter 3 in [2].

Theorem 3.0.5. *Let $a_i(x)$, where $i = 0, \ldots, n$ and $\phi(x)$, all be continuous complex-valued functions defined on an interval I, with $a_n(x)$ never 0, and for a C^∞ function f on I, let*

$$Df = \sum_{i=n}^{0} a_i(x) \frac{d^i f}{dx^i} + \phi(x).$$

We consider on I the non-homogeneous ODE $Df = 0$ and also the homogeneous equation $\Delta(f) = 0$, where $\Delta = D - \phi$. Then, the following hold:

1. *Let $c_n, \ldots c_0$ be $n + 1$ given complex numbers and x_0 be fixed in I. Then there is a unique function $y(x)$ satisfying the initial conditions, $y^i(x_0) = \frac{d^i f}{dx^i}(x_0)$ for $i = 0, \ldots n$ and the non-homogeneous ODE,*

$$Df = 0.$$

2. *The homogeneous equation has $n+1$ linearly independent solutions over $\mathbb{C}$.*
3. *The $\mathbb{C}$ linear span of these form an $n+1$-dimensional vector space V over $\mathbb{C}$ (with the operations of adding functions pointwise and multiplying functions pointwise by complex numbers) and comprise all solutions of the homogeneous equation.*
4. *If $y(x)$ is a solution to the non-homogeneous equation, then $y(x) + v(x)$, where $v \in V$ comprise all solutions to the homogeneous equation.*

Exercise 3.0.6. Show that the solution to any nth-order differential equation (with variable coefficients) can be reduced to the solution of a system of n first-order differential equations (with variable coefficients).

Exercise 3.0.7. In connection with our above remarks on second-order equations try using the techniques established to solve the second-order homogeneous equation with constant real coefficients, $a \neq 0$, b and c,

$$a\frac{d^2 f}{dx^2} + b\frac{df}{dx} + cf = 0,$$

where we make no assumptions concerning the roots of the associated quadratic equation, $aX^2 + bX + c = 0$.

Actually, this slightly more general, second-order differential equations can be used to solve other mechanics problems such as *forced oscillations* of a spring, with *damping* such as in the following exercise.

Exercise 3.0.8. Consider the problem of the harmonic oscillator, but now with damped vibrations, namely, at any given time, the force F_d of damping occurs proportional to the velocity, i.e., $F_d = rv$, where r is a constant. Thus, if $x(t)$ is the position at time t,

$$\frac{d^2 x}{dt^2} + r\frac{dx}{dt} + k^2 x = 0.$$

Then,

$$x(t) = Ae^{\frac{rt}{2}}\cos(\omega t) + Be^{\frac{rt}{2}}\sin(\omega t),$$

where $\omega = \sqrt{k^2 - (\frac{r}{2})^2}$. Note when $r = 0$, as expected, we get $\omega = k$ and $x(t) = A\cos(\omega t) + B\sin(\omega t)$.

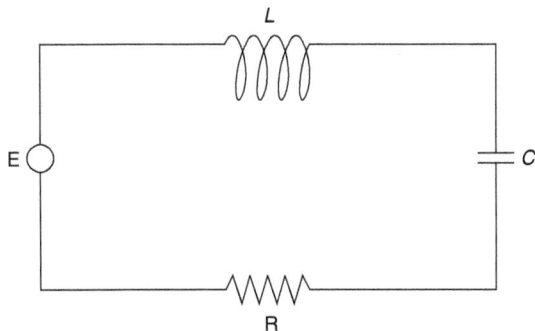

Figure 3.1. An electric circuit containing a resistor, coil and capacitor connected in series.

Characteristic of mathematics, these same principles apply equally well to a seemingly unrelated situation such as current flowing in an electric circuit. In more detail, consider an electric circuit containing a resistor R, a coil L, a capacitor C and a source $E(t)$ of electrical current at time t called a generator. We let $i(t)$ denote the current and $q(t)$ the charge at time t (all of which we shall assume to be smooth). Assuming that the wire itself has no resistance, the voltage drop across the resistor is $Ri(t)$ and across the capacitor is $\frac{1}{C}q(t)$, where $q(t)$ is the charge and C is the capacitance of the capacitor and the voltage drop across the coil is proportional to the rate of change of $i(t)$, so is $Li'(t)$. Kirchoff's (1824–1877) law tells us that the sum of these voltage drops is $E(t)$. Thus, we get the following ODE:

$$Li'(t) + Ri(t) + \frac{q(t)}{C}q(t) = E(t).$$

As $i(t) = q'(t)$, differentiating gives

$$Li''(t) + Ri'(t) + \frac{1}{C}i(t) = E'(t).$$

Here, $E(t)$ is usually some type of bounded periodic function. This gives a non-homogeneous second-order ODE in the current, $i(t)$, with principal part having constant coefficients, which can be readily solved.

Chapter 4

Some More Advanced Topics in ODEs

In this chapter, we turn to some second-order ODEs, but now with variable coefficients. This will bring us into contact with orthogonal families of special functions, the orthogonality not being with respect to the usual measure, but rather with respect to a weighted multiple of it. We shall limit ourselves to Hermite functions, but there are many others, such as Bessel, Legendre and Laguerre functions. These comprise a whole subject called *orthogonal polynomials*.

A word about terminology. Here, we are interested in a weight function ω which is defined on all of $\mathbb{R}$, taking non-negative values, and which has finite integral. $\int_{-\infty}^{\infty} \omega(x)dx < \infty$.

In general, letting $a < b$, we will be considering an "interval" I to be one of the following: $[a, b]$, (a, ∞), $(a, b]$, (a, b), $[a, b)$, $(-\infty, b)$ or $(-\infty, \infty) = \mathbb{R}$. Therefore, on an interval, we get a "weighted measure" $\omega(x)dx$ and ω itself is called a "weight function". The purpose of this is to compute integrals, $\int_I f(x)\omega(x)dx$, where f is a continuous function on I (or more generally a measurable function) and dx stands for the Riemann integral (or more generally Lebesgue measure). However, when I is not finite, we shall have to impose conditions on the integrand to insure that its integral is never the less finite.

In this context, two such real-valued functions f and g on I are said to be orthogonal if $\int_I f(x)g(x)\omega(x)dx = 0$, and a family f_n of such functions are called orthonormal when $\int_I f_m(x)f_n(x)\omega(x)dx = \delta_{m.n}$. This last term is called the Kronecker delta. As we shall see, the Hermite functions will do all of this when $I = (-\infty, \infty)$ and ω is the Gaussian distribution (see Appendix A).

4.1 The Method of Frobenius: Second-Order Equations with Variable Coefficients

Concerning properties of real analytic functions, the reader should consult Appendix D.

Earlier, we noted that a second-order homogeneous linear differential equation with variable coefficients is of the form

$$a(x)y'' + b(x)y' + c(x)y = 0,$$

where the coefficients $a(x)$, $b(x)$ and $c(x)$ are continuous real-valued functions. In this section, we will study one of these, namely, the *Hermite equation* using the method of Frobenius.[1] Its solutions involve the so-called *Hermite polynomials*, H_n, $(n = 0, 1, 2, \ldots)$.

Here, we shall assume that the coefficient functions $a(x)$, $b(x)$ and $c(x)$ are all defined and real analytic on some interval $J = (\alpha, \beta)$ with $a(x) \neq 0$ everywhere on (α, β). In fact, in many cases, $a(x)$, $b(x)$ and $c(x)$ will be polynomials on $(-\infty, \infty)$ or an open interval (α, β) in $\mathbb{R}$.

As we observed, since $a(x) \neq 0$ everywhere on J, we can divide by it while maintaining our conditions on the other coefficients and write

$$y'' + p(x)y' + q(x)y = 0, \tag{4.1}$$

where $p(x) = \frac{b(x)}{a(x)}$ and $q(x) = \frac{c(x)}{a(x)}$ are themselves real analytic on (α, β) as well.[2]

---

[1]Frobenius (1849–1917) was a German mathematician best known for his contributions to the theory of elliptic functions, differential equations, number theory, real division algebras and group representations.

[2]To see this, one extends both $a(x)$ and $b(x)$ as analytic functions into the complex domain (in a disk centered at the midpoint of J and of radius $\frac{1}{2}$ the length of (α, β)). Then by continuity, making the disk, say D smaller if necessary to make sure that $a(z) \neq 0$ in this subdisk. Now, we have two holomorphic functions $a(z)$ and $b(z)$ defined on D with $a(z)$ never vanishing. By the usual differentiation formula for a quotient, we see that $\frac{b(z)}{a(z)}$ is holomorphic and therefore analytic. Hence, so is its restriction, $\frac{b(x)}{a(x)}$ (see, e.g., [9]).

As a result of this, our next theorem will show that the solutions of equation (4.1) are also analytic on $(\alpha, \beta) = J$. We shall also see that, just as in the case of second-order constant coefficient equations, the solutions depend on two arbitrary constants, a_0 and a_1.

Theorem 4.1.1. *Let $x_0 \in (\alpha, \beta)$ and a_0 and a_1 be arbitrary constants. Then there is a unique solution $y(x)$ to equation (4.1) in some neighborhood of x_0 within J which is real analytic (i.e., is given by a convergent power series on (α, β)) and satisfies the initial conditions $y(x_0) = a_0$ and $y'(x_0) = a_1$.*

Proof. By translation, we may assume $x_0 = 0$. This will make our notation simpler, so instead of power series in $x - x_0$, we just have to consider power series in x. Let $p(x) = \sum_{n=0}^{\infty} p_n x^n$ and $q(x) = \sum_{n=0}^{\infty} q_n x^n$ converge for $|x| < r$, where $r > 0$. If $y(x)$ is supposed to end up a power series about 0, then we write

$$y(x) = \sum_{n=0}^{\infty} a_n x^n,$$

assuming it also converges for $|x| < r$, and try to find what requirements the a_n have to satisfy. Differentiating term by term and using multiplying and adding convergent power series substituting into equation (4.1) (see Appendix D), we get

$$\sum_{n=0}^{\infty} \left[(n+1)(n+2)a_{n+2} + \sum_{n=0}^{n} p_{n-k}(k+1)a_{k+1} \right.$$

$$\left. + \sum_{n=0}^{n} q_{n-k}a_k \right] x^n = 0.$$

Now, since a power series which is identically zero must have all its coefficients zero (and conversely) (for all this, see Appendix D), we obtain the recursion formula,

$$(n+1)(n+2)a_{n+2} = -\sum_{n=0}^{n} [(k+1)p_{n-k}a_{k+1} + q_{n-k}a_k], \quad (n \geq 0).$$

$$(4.2)$$

To get a better feel for this, let us write out the first few of these equations,

- $2a_2 = -(p_0a_1 + q_0a_0)$.
- $2 \cdot 3a_3 = -(p_1a_1 + 2p_0a_2 + q_1a_0 + q_0a_1)$.
- $3 \cdot 4a_4 = -(p_2a_1 + 2p_1a_2 + 3p_0a_3 + q_2a_0 + q_1a_1 + q_0a_2)$.

These formulas determine a_n for $n \geq 2$ in terms of a_0 and a_1. For instance, the first formula determines a_2. Having done so, the second formula determines a_3, etc., and they do so in such a way that the resulting formal series satisfies equation (4.1). Furthermore, from the formulas $y(x) = \sum_{n=0}^{\infty} a_n x^n$ and $y'(x) = \sum_{n=0}^{\infty}(n+1)a_{n+1}x^n$, it follows immediately that $y(0) = a_0$ and $y'(0) = a_1$. It only remains to show the series for $y(x)$ converges in J.

We now complete the proof of Theorem 4.1.1 by showing that y is analytic. Let $0 < r_0 < r$. Since the series for $p(x)$ and $q(x)$ converge for $x = r_0$, there is a constant M for which both $|p_n r_0^n| \leq M$ and $|q_n r_0^n| \leq M$ for all $n \geq 0$. Hence,

$$(n+1)(n+2)|a_{n+2}| \leq \frac{M}{r_0^n} \sum_{k=0}^{n}[(k+1)|a_{k+1}| + |a_k|]r_0^k.$$

Throwing in $M|a_{n+1}|r$ which can only make the inequality more so, we get

$$(n+1)(n+2)|a_{n+2}| \leq \frac{M}{r_0^n} \sum_{k=0}^{n}[(k+1)|a_{k+1}| + |a_k|]\,r_0^k + M|a_{n+1}|r_0.$$

For $n \geq 0$, we define a sequence of non-negative terms b_n by $b_0 = |a_0|$, $b_1 = |a_1|$ and for $n \geq 2$ by

$$(n+1)(n+2)b_{n+2} = \frac{M}{r_0^n} \sum_{k=0}^{n}[(k+1)b_{k+1} + b_k]r_0^k + Mb_{n+1}r_0.$$

$$(4.3)$$

Evidently, $0 \leq |a_n| \leq b_n$ for all n. The question is then: For which x does the series $\sum_{n=0}^{\infty} b_n x^n$ converge? Equation (4.2) tells us that

$$n(n+1)b_{n+1} = \frac{M}{r_0^{n-1}} \sum_{k=0}^{n-1}[(k+1)b_{k+1} + b_k]r_0^k + Mb_n r_0$$

and

$$(n-1)(n)b_n = \frac{M}{r_0^{n-2}} \sum_{k=0}^{n-2}[(k+1)b_{k+1} + b_k]r_0^k + Mb_{n-1}r_0.$$

Multiplying the first of these by r_0 and substituting the second into it yields $r_0 n(n+1)b_{n+1} = \frac{M}{r_0^{n-2}} \sum_{k=0}^{n-2}[(k+1)b_{k+1}+b_k]r_0^k + r_0 M(nb_n + b_{n-1}) + Mb_n r_0^2$. But this is $[(n-1)n + r_0 Mn + Mr_0^2]b_n$. Thus,

$$\frac{b_{n+1}}{b_n} = \frac{(n-1)n + r_0 Mn + Mr_0^2}{r_0 n(n+1)}.$$

From this, we see easily that

$$\lim_{n\to\infty} \frac{b_{n+1}}{b_n} = \frac{1}{r_0}$$

and hence that

$$\lim_{n\to\infty} \frac{b_{n+1}x^{n+1}}{b_n x^n} = \frac{|x|}{r_0}.$$

The ratio test tells us that the series $\sum_{n=0}^{\infty} b_n x^n$ converges for $|x| < r_0$ and the comparison test tells us that the same is true for $\sum_{n=0}^{\infty} a_n x^n$. Since this is so for all $r_0 < r$, we conclude $\sum_{n=0}^{\infty} a_n x^n$ converges whenever $|x| < r$. $\square$

Exercise 4.1.2. Apply the method of Frobenius to solve the *harmonic oscillator* equation $\frac{d^2y}{dx^2} + \omega^2 y = 0$, where ω is a real parameter. Do the same with the *harmonic oscillator with drag* $\frac{d^2y}{dx^2} - k\frac{dy}{dx} + \omega^2 y = 0$, where k is a positive constant. How do these solutions compare with those we obtained earlier?

Exercise 4.1.3. We now turn briefly to the Riccati[3] equation. This is a first-order differential equation of the form

$$y' = c(x) + b(x)y + a(x)y^2, \tag{4.4}$$

where the coefficients are assumed to be smooth functions on some interval. This equation is a natural extension of the first-order linear

---

[3] J. Riccati (1676–1754) formulated and solved the Riccati equation.

equation, as it reduces to a linear equation if the coefficient $a(x) \equiv 0$. Of course, if the coefficients a, b and c are constants, then the Riccati equation allows the separation of variables and the general solution is

$$\int \frac{1}{ay^2 + by + c} dy = C - x.$$

We ask the reader to check this. In general, the Riccati equation cannot be solved by elementary methods. However, if a particular solution $y_*(x)$ is known, then the general solution has the form

$$y(x) = y_*(x) + z(x),$$

where $z(x)$ is the general solution of the *Bernoulli equation*,

$$z' - (b - 2ay_*)z = az^2.$$

To see this, set $y(x) = y_*(x) + z(x)$, where $z(x)$ is a new unknown function. Substituting $y = y_* + z$ into the Riccati equation, we find that

$$y'_* + z' = c(x) + b(x)(y_* + z) + a(x)(y_*^2 + 2y_*z + z^2).$$

Since $y_*(x)$ is a solution, we get $z' = b(x)z + a(x)(2y_*z + z^2)$, or

$$z' - [b(x) + 2a(x)y_*(x)]z = a(x)z^2. \tag{4.5}$$

This is called the Bernoulli equation, which with the substitution $w = \frac{1}{z}$ reduces to a linear equation. We ask the reader to check this. The reader should also observe that this last maneuver was what we used in solving the two-body problem.

More generally, we consider the following equation, which is also called the Bernoulli equation:

$$\frac{dy}{dx} + py = qy^n.$$

Here, n is an integer ≥ 2 and p and q are smooth functions of x. Letting $z = y^{1-n}$, one sees easily that

$$\frac{dz}{dx} + (1 - n)pz = (1 - n)q$$

and so can be reduced to a linear equation.

4.2 The Hermite Equation

Here, we solve the differential equation called the Hermite[4] equation:

$$y'' - 2xy' + \lambda y = 0, \quad -\infty < x < \infty, \tag{4.6}$$

where $\lambda \in \mathbb{R}$ is a parameter. Since in equation (4.6) $a(x) \equiv 1$, the question of quotients of analytic functions doesn't arise at all. Here, the interval $(\alpha, \beta) = \mathbb{R}$.

Solution. Theorem 4.1.1 tells us that all solutions are given by power series

$$y(x) = \sum_{n=0}^{\infty} a_n x^n,$$

with $a_0 = y(0)$ and $a_1 = y'(0)$ arbitrary, and this series converges for $|x| < \infty$. We shall also shortly impose certain *boundary conditions* on the solutions.

Note that if $y(x)$ is a solution to Hermite's equation, then so is $y(-x)$. Hence, $y(x) + y(-x)$ and $y(x) - y(-x)$ are also solutions, so we can concern ourselves only with *even* and *odd* solutions. This means we consider power series of the form $y(x) = \sum_{n=0}^{\infty} a_{2n} x^{2n}$ or $y(x) = \sum_{n=0}^{\infty} a_{2n+1} x^{2n+1}$. The recursion formulas then say

$$\frac{a_{n+2}}{a_n} = \frac{2n - \lambda}{(n+1)(n+2)}. \tag{4.7}$$

The boundary conditions that we alluded to earlier are that $y(x) = O(x^k)$ as $x \to \pm\infty$ for a certain finite power k.[5] So, taking either an even or odd solution, we see that these boundary conditions force $\lambda = 2n$ for some n. For otherwise, if $a_0 \neq 0$ (respectively $a_1 \neq 0$), then in either case, we have an infinite series and $y(x)$ cannot be $O(x^k)$. Of course, if either $a_0 = 0$ (respectively $a_1 = 0$), then the even (respectively odd) function is identically zero. Thus, we know $\lambda = 2, n$ and therefore, our solutions, satisfying the boundary conditions, are either *polynomials consisting of all even degree terms*

---

[4]Charles Hermite (1822–1901), a 19th century mathematician who made many contributions to number theory, quadratic forms, analysis and algebra. In 1873, he proved the transcendence of e.

[5]$y(x) = O(x^k)$ as $x \to \pm\infty$ means $\frac{y(x)}{x^k}$ is bounded near $\pm\infty$.

or those consisting of all odd degree terms! These are called *Hermite polynomials* and the equation is actually

$$y'' - 2xy' + 2ny = 0.$$

The recursion formula tells us these polynomials are given by

$$H_n(x) = (2x)^n - \frac{n(n-1)}{1!}(2x)^{n-2}$$
$$+ \frac{n(n-1)(n-2)(n-3)}{2!}(2x)^{n-4} - \cdots,$$

where n is either odd or even and the last term is either $(-1)^{\frac{n}{2}}\frac{n!}{\frac{n}{2}!}$, if n is even, or $(-1)^{\frac{n-1}{2}}\frac{n!}{\frac{n-1}{2}!}2x$, if n is odd. In other words,

$$H_n(x) = \sum_{k=0}^{[\frac{n}{2}]}(-1)^k\frac{n!}{k!(n-2k)!}(2x)^{n-2k}, \qquad (4.8)$$

where $[\frac{n}{2}]$ means the greatest integer $\leq \frac{n}{2}$.

Thus, for example,

$$H_0(x) = 1, \ H_1(x) = 2x, \ H_2(x) = 4x^2 - 2, \ H_3(x) = 8x^3 - 12x,$$
$$H_4(x) = 16x^4 - 48x^2 + 12, \ H_5(x) = 32x^5 - 16x^3 + 120x, \ldots$$

Remark 4.2.1. The function $f(x) = e^{-x^2}$ has much to do with the Hermite equation and its solutions. In fact, the reader can easily check that Hermite's equation is equivalent with the equation

$$(e^{-x^2}y')' + \lambda e^{-x^2}y = 0. \qquad (4.9)$$

This is the *Sturmian*[6] form of the Hermite equation.

Thus, Hermite's polynomials satisfy

$$H_n''(x) - 2xH_n'(x) + 2nH_n(x) = 0 \qquad (4.10)$$

---

[6] Jacques Charles Francois Sturm (1803–1855) was a Swiss mathematician whose main work was in what is now called the *Sturm–Liouville theory* of differential equations.

and

$$(e^{-x^2} H_n'(x))' + 2ne^{-x^2} H_n(x) = 0. \qquad (4.11)$$

Corollary 4.2.2 (Orthogonality of the Hermite polynomials).
The Hermite polynomials are orthogonal with respect to the weight e^{-x^2} (Recall this is the Gaussian distribution function we observed earlier (see Appendix A) on the interval $(-\infty, \infty)$, i.e., for $m \neq n$,

$$\int_{-\infty}^{\infty} H_n(x) H_m(x) e^{-x^2} dx = 0.$$

Proof. Let $m \neq n$. Multiplying 4.11 by H_m, we get

$$(e^{-x^2} H_n'(x))' H_m(x) + 2ne^{-x^2} H_n(x) H_m(x) = 0,$$

which is symmetric with respect to n and m. So, we also have

$$(e^{-x^2} H_m'(x))' H_n(x) + 2me^{-x^2} H_m(x) H_n(x) = 0.$$

Subtracting the second of these equations from the first, we get

$$\left[e^{-x^2} (H_m H_n' - H_n H_m') \right]' + 2(n - m)e^{-x^2} H_n(x) H_m(x) = 0.$$

Now, integrating from $-\infty$ to ∞,

$$\left[e^{-x^2} \left(H_m(x) H_n'(x) - H_n(x) H_m'(x) \right) \right]_{-\infty}^{\infty}$$

$$+ 2(n - m) \int_{-\infty}^{\infty} e^{-x^2} H_n(x) H_m(x) = 0$$

$$= 2(n - m) \int_{-\infty}^{\infty} e^{-x^2} H_n(x) H_m(x) = 0.$$

$\square$

To reiterate, Hermite's polynomials are an example of the wider class of *special functions* called *orthogonal polynomials.*[7] Orthogonal polynomials can also be defined as the coefficients in expansions in powers of t of suitably chosen functions $\psi(x,t)$ called *generating functions*.

Definition 4.2.3. A *generating function* for H_n is a function $\psi(x,t)$, satisfying

$$\psi(x,t) = \sum_{n=0}^{\infty} H_n(x)\frac{t^n}{n!}, \quad |t| < \infty.$$

We shall now find a generating function for H_n.

Proposition 4.2.4.

$$\psi(x,t) = e^{-t^2+2tx} = \sum_{n=0}^{\infty} H_n(x)\frac{t^n}{n!}, \quad |t| < \infty.$$

Proof.   Using equation (4.8), we have

$$\sum_{n=0}^{\infty} H_n(x)\frac{t^n}{n!} = \sum_{n=0}^{\infty}\left[\sum_{k=0}^{[\frac{n}{2}]}(-1)^k\frac{1}{k!(n-2k)!}(2x)^{n-2k}\right]t^n$$

$$= \left[\sum_{n=0}^{\infty}\frac{(-1)^n}{n!}t^{2n}\right]\left[\sum_{n=0}^{\infty}\frac{(2xt)^n}{n!}\right]$$

$$= e^{-t^2}e^{2xt} = e^{-t^2+2xt} = \psi(x,t). \qquad \square$$

---

[7]A system of real functions $\{\varphi_n(x)\}$, $n = 0, 1, 2, \ldots$ is called *orthogonal with weight* $w(x) > 0$ on the interval $[a,b]$ if

$$\int_a^b \varphi_n(x)\varphi_m(x)w(x)dx = 0,$$

for every $n \neq m$, where $[a,b]$ could be a finite interval, a ray or $\mathbb{R}$. The weight function for Hermite's polynomials is $w(x) = e^{-x^2}$ and the interval $(-\infty, \infty)$. Other examples of such polynomials are the Legendre, Laguerre, Chebyshev, Jacobi and Gegenbauer polynomials. In addition to the orthogonality property, these polynomials are solutions of differential equations of a simple form and have other general properties with many applications in mathematical physics, approximation theory, partial differential equations, engineering, etc.

As a consequence, we get *Rodrigues' formula.*[8] It gives an efficient way to generate the Hermite polynomials.

Corollary 4.2.5.

$$H_n(x) = (-1)^n e^{x^2} \frac{d^n}{dx^n} e^{-x^2}, \quad n = 0, 1, 2, 3, \ldots \tag{4.12}$$

Proof. From $\psi(x,t) = \sum_{n=0}^{\infty} H_n(x) \frac{t^n}{n!}$ and upon differentiating the power series term by term (see Appendix D), we get

$$\left. \frac{\partial^n \psi(x,t)}{\partial t^n} \right|_{t=0} = H_n(x).$$

On the other hand, by induction, one can easily show that

$$\left. \frac{\partial^n \psi(x,t)}{\partial t^n} \right|_{t=0} = \left. \frac{\partial^n}{\partial t^n} \left(e^{x^2} e^{-(t-x)^2} \right) \right|_{t=0} = (-1)^n e^{x^2} \frac{d^n}{dx^n} e^{-x^2}. \qquad \square$$

The following useful formula gives the derivative of a Hermite polynomial recursively in terms of the next lower Hermite polynomial.

Corollary 4.2.6. *For* $n \geq 1$, $H_n'(x) = 2nH_{n-1}(x)$.

Proof. This follows from $\frac{\partial \psi(x,t)}{\partial x} = 2t\psi(x,t)$. $\qquad \square$

Corollary 4.2.7. *For* $n \geq 1$, $H_{n+1}(x) - 2xH_n(x) + 2nH_{n-1}(x) = 0$.

Proof. This follows from $\frac{\partial \psi(x,t)}{\partial t} = 2(x-t)\psi(x,t)$. $\qquad \square$

Final remark: By some miracle of fate, the Hermite equation and the resultant *Hermite functions,*

$$h_n(x) = e^{\frac{-x^2}{2}} H_n(x),$$

---

[8]Benjamin Olinde Rodrigues (1794–1851) was from Sephardic Jewish of Portugese ancestry from Bordeaux. As a banker, he supported Claude Henri Saint-Simon (the founder of socialism) in his destitute old age and became one of his earliest disciples. He discovered the above formula in 1816, but soon thereafter became interested in the scientific organization of society and never returned to mathematics.

turn out to have a significance well beyond what Hermite envisaged in the mid-19th century. For three quarters of a century later, Hermite functions played an important role in solving the Schrödinger equation of quantum mechanics which is the quantum mechanical analog of the harmonic oscillator of classical mechanics. This is

$$\frac{d^2\psi}{dx^2} + \left(\frac{2E}{h\nu} - x^2\right)\psi = 0,$$

where E is the the total energy, ν is the frequency, h is Planck's constant and ψ is an unknown real-valued function of x, the position. There are two additional conditions to be satisfied:

1. ψ vanishes at infinity. This means that, given any $\epsilon > 0$, there is always a closed interval about 0, outside of which $|\psi(x)| \leq \epsilon$.
2. $\int_{-\infty}^{+\infty} |\psi|^2 dx = 2\pi\sqrt{(\frac{\nu m}{h})}$, where m is the mass.

Leaving aside these two conditions, with a change of notation, this second-order differential equation is just the Hermite equation. Hence, we know that it has solutions satisfying $E = h\nu(n + \frac{1}{2})$ for some non-negative integer n. Moreover, the solutions are the Hermite functions $\psi(x) = ce^{-x^2}2H_n(x)$, where c is a constant and H_n is the nth Hermite polynomial. Of course, the first condition is automatically satisfied (prove), while the second forces $c = \frac{4\pi\nu m}{2^{2n}(n!)^2 h}^{\frac{1}{4}}$.[9]

---

[9]Erwin Schrödinger (1887–1961) was an Austrian physicist who developed fundamental results in quantum theory and contributed to statistical mechanics, thermodynamics, electrodynamics and general relativity.

Chapter 5

Approximation of Solutions

5.1 Small Vibrations of a Pendulum

In this chapter, we deal with the classical problem of the motion of a pendulum undergoing small vibrations. This means that a point mass at the end of a weightless rod or string of length l and weight $W = mg$. Let θ be the angle between the rod and the vertical. We resolve the vector W into its normal and tangential components. Since $\frac{ds}{dt} = l\frac{d\theta}{dt}$, and therefore, $\frac{d^2s}{dt^2} = l\frac{d^2\theta}{dt^2}$, and the tangential component of the weight vector is $W\sin(\theta)$ pointing downward, $F = ma$ tells us that $ml\frac{d^2\theta}{dt^2} = -mg\sin(\theta)$, where as usual m cancels, so the motion is independent of mass. Thus,

$$\frac{d^2\theta}{dt^2} + \frac{g}{l}\sin(\theta) = 0.$$

Now, small vibrations means that the angle θ is small. Since as θ approaches zero, $\lim \frac{\sin(\theta)}{\theta} = 1$, we boldly replace $\sin(\theta)$ by θ itself (otherwise, we would not be able to solve the ODE as we had done in earlier chapters). Thus,

$$\frac{d^2\theta}{dt^2} + \frac{g}{l}\theta = 0.$$

Since $\frac{g}{l} > 0$, we are again back in the harmonic oscillator problem. Hence,

$$\theta(t) = A\sin\left(\sqrt{\frac{g}{l}}t\right) + B\cos\left(\sqrt{\frac{g}{l}}t\right),$$

53

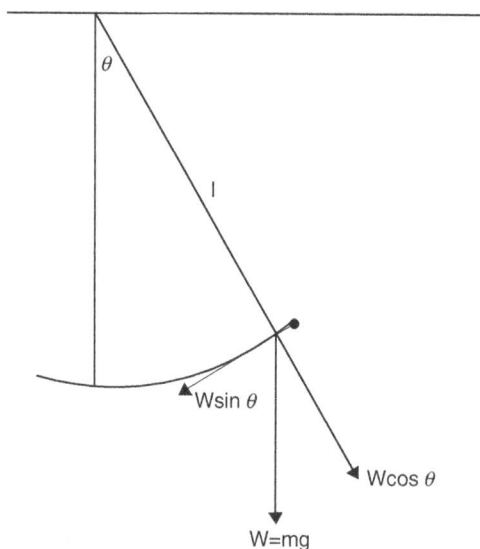

Figure 5.1.   The gravitational forces on a pendulum.

and the period, T, is (approximately) $\frac{2\pi}{\sqrt{\frac{g}{l}}}$. Thus, we recapture

Galileo's discovery that, under small vibrations, $\frac{T}{\sqrt{l}}$ is constant, i.e.,

T is proportional to $\sqrt{l}$ with a known constant of proportionality.

As we have been using the concepts of normal and tangential components of a vector, here is another application of these ideas which enables us to analyze the forces that move a sailboat through the water.

To do so, we can neglect any effect of the wind on the hull since this force is negligible compared to the force of the wind on the sail. (For simplicity, we shall assume there is only one sail.) Moreover, although the wind acts on every point of the sail, we can consider it to be acting at a single point, namely, the center of gravity of the sail with an equivalent force so that we are dealing with forces (vectors) acting at a point. This enables us to use the *parallelogram law* for addition of vectors, which goes as follows: Let v and w be *linearly independent* vectors in a vector space (implying that they are each non-zero and point in different directions and so span a plane). In this plane, consider the parallelogram generated by v and w. In geometric form, the *parallelogram law* states that the vector $v+w$ is represented in both *magnitude* and *direction* by the diagonal of the parallelogram formed by v and w. An important special case of this is when the non-zero v and w happen to be perpendicular. Then they are indeed linearly independent. Consider a vector u and a coordinate system with an x- and y-axis in this plane. One can resolve u (write it) as the sum of the vectors u_x and u_y, where these are the components of u, namely, the projections of u onto the x and y axes, respectively. Thus, u is the vector sum of its components, $u = u_x + u_y$.

Now, let us consider the effect on a sailboat being blown by the wind with a force u. Resolving this vector into its tangential and orthogonal components with respect to the sail, $u = u_{\text{tang}} + u_\perp$. Since u_{tang} will have no effect on the sail, and therefore, the boat we can ignore it. Thus, the effect of the force u is the same as that of $u_\perp$. What is this effect? Resolving $u_\perp$ into the sum of its component, u_{axis} along the axis of the boat and u_c, its component along the perpendicular axis yields $u_\perp = u_{\text{axis}} + u_c$, the effect of u_c being canceled by the boat's centerboard (or keel). Thus, the force that acts to drive the boat directly forward is u_{axis}.

5.2 Predators and Prey

We now turn to the *predator–prey problem* which was first formulated and solved by Volterra.[1]

First, the heuristics. Imagine an island whose only inhabitants are foxes and rabbits. The foxes like to eat the rabbits, who themselves eat the unlimited quantities of grass on the island. When there are lots of rabbits available, the foxes do well and multiply. However, this multiplying perforce decreases the number of rabbits available. As a result, some of the foxes become weaker and are not so good at catching rabbits; thus, they die and the rabbits multiply. As a result, the foxes begin to do better and the rabbits worse, etc.

---

[1]Vito Volterra (1860–1940) was an Italian Jewish mathematician who strongly influenced the modern development of integral and differential equations and their applications to mathematical physics and mathematical biology. Volterra opposed fascism from the outset, refused to take the required loyalty oath to the government of Mussolini and as a result was fired by the University of Rome. Shortly after his death, the Italian police came to arrest him, but they were too late.

Let $x(t)$ and $y(t)$ be the number of rabbits and foxes, respectively, at time t. Since there is plenty of grass, if there were no foxes, we can assume as above that the normal rate of increase of the rabbits is given by $x'(t) = ax(t)$, where $a > 0$ is some positive constant. However, with foxes present, this decreases proportionally to the number of encounters of the two. Thus, actually, $x' = ax - bxy$, where $b > 0$ is the constant giving the proportion of these encounters, which have been successful from the foxes' point of view. Similarly, $y' = -cy + dxy$, where $c, d > 0$. Thus, Volterra's predator–prey equations are

$$\frac{dx}{dt} = x(a - by)$$

and

$$\frac{dy}{dt} = -y(c - dx).$$

These are a set of two simultaneous first-order equations, but because of the xy terms, they are nonlinear and can't be solved in closed form the way we have solved earlier ODEs. Letting t vary and considering $(x(t), y(t))$ as a curve in the plane, we can eliminate t and look at the curve in its (x, y) coordinates. In terms of differential forms, we get

$$\frac{(a - by)dy}{y} = -\frac{(c - dx)dx}{x}.$$

Note that when $x = \frac{a}{b}$ and $y = \frac{c}{d}$, this equation holds, i.e., the point $(\frac{a}{b}, \frac{c}{d})$ lies on the curve.

Upon integrating, the equation above yields

$$a \log(y) - by = -c \log x + dx + C,$$

where $C = \log K$ is a constant of integration (the log function maps onto $\mathbb{R}$). This means that

$$y'ae^{-by} = Kx'-ce^{dx}.$$

Another way to view this is $a \log(y) + c \log(x)$ must lie on one of a family of parallel lines, $dx + by + C = 0$.

Now, it would be nice to solve for y in terms of x or vice versa. Unfortunately, that can't be done either. The only thing we can solve

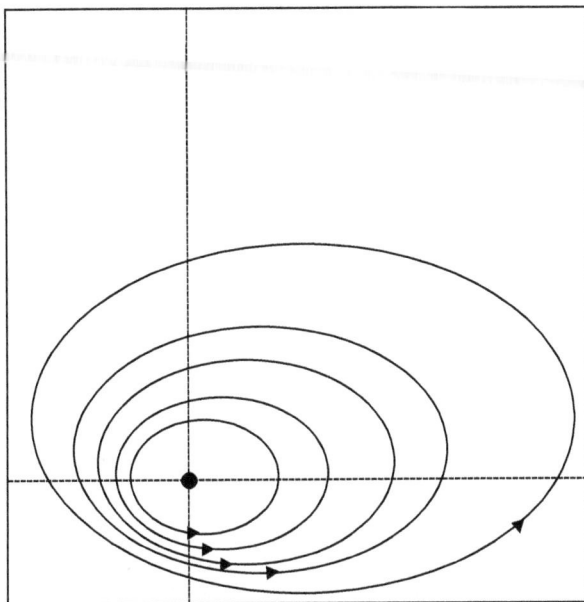

Figure 5.2. The qualitative solutions to a predator prey problem.

for is K in terms of initial values of x and y and this doesn't help. Taking a few numerical values at random suggests that this curve is some kind of simple closed curve in the first quadrant as in Figure 5.2. (The reader should verify this.) Note that the smaller inner curve is more elliptical than the outer one.

What to do? First, let us find the places in the plane where both $-y(c-dx)$ and $x(a-by)$ are zero. Thus, we are looking at the places where a vector field in $\mathbb{R}^2$ is zero. Evidently, in our problem, we can assume neither x nor y is zero, since they are both positive. Hence, the unique point where this vector field vanishes is $(\frac{c}{d}, \frac{a}{b})$. Let us translate the curve so that the point $(\frac{c}{d}, \frac{a}{b})$ lies at the origin. Thus, we take $x = \frac{c}{d} + X$ and $y = \frac{a}{b} + Y$ and ask: What does this curve looks like in the new X, Y coordinates? Since $\frac{dX}{dt} = \frac{dx}{dt}$ and $\frac{dY}{dt} = \frac{dy}{dt}$,

$$\frac{dX}{dt} = -\frac{bc}{d}Y - bXY$$

and

$$\frac{dY}{dt} = -\frac{ad}{b}X + dXY.$$

Now, for small X and Y, XY is much smaller and perhaps could be disregarded! In that case, since b and d are constants, for (x, y) near $\left(\frac{c}{d}, \frac{a}{b}\right)$, we would have

$$\frac{dX}{dt} = -\frac{bc}{d}Y$$

and

$$\frac{dY}{dt} = -\frac{ad}{b}X.$$

Here, we have now have a system of two *linear* homogeneous ODEs with constant coefficients!

Dividing yields $\frac{dY}{dX} = -\frac{ad^2}{b^2c}\frac{X}{Y}$ or $b^2cY\,dY + a^2dX\,dX = 0$. Integrating tells us $\frac{b^2c}{2}Y^2 + \frac{a^2d}{2}X^2 = D$ (where $D > 0$ since the left side is certainly positive). Thus, we get a family of ellipses $a^2dX^2 + b^2cY^2 = 2D$ centered at the origin. This means the untranslated curve is an ellipse centered at $\left(\frac{c}{d}, \frac{a}{b}\right)$ whose eccentricity is determined by the constants a, b, c, d and we can hope that the original curve is some kind of simple closed, probably convex curve with $\left(\frac{c}{d}, \frac{a}{b}\right)$ in its interior.

The reader might wish to solve the pendulum problem and the system of predator–prey equations without approximations, i.e., solving the original unadjusted system. One way of accomplishing this is by use of Liapunov stability.

5.3   Liapunov Stability

Although here all we need to consider is a 2×2 autonomous dynamical system for either the pendulum or the predator–prey problem, it takes no more effort to do the Liapunov[2] theory in n dimensions, which would be useful in other applications.

So, we consider an autonomous dynamical system in dimension n,

$$\frac{dX}{dt} = F(X),$$

where F is a C^1 function on a domain D in $\mathbb{R}^n$. Actually, we shall have to make slightly stronger assumptions about F as we go. This is

---

[2]Alexandr Liapunov (1857–1918) was a Russian mathematician and physicist.

the general formulation of an autonomous dynamical system which in the 2 × 2 case would be specified as follows:

$$\frac{dx}{dt} = F(x, y)$$

and

$$\frac{dy}{dt} = G(x, y),$$

where F and G are C^1 functions on an open domain D in the x, y plane.

The reader should bear in mind the following:

1. The performance of devices are often governed by a system of ODEs, and as we know, such a system always has an infinite number of solutions. However, even if we could find and then solve the ODE governing a device such as a clock, or an electric circuit, when started, it is practically impossible to determine initial conditions to single out a solution. Moreover, it is not even desirable to do so. This is because after the device operations stabilize, the operating characteristics rather quickly also stabilize. In this way, the ODE which describes the operation of the device will have stationary solutions which we can endeavor to find.

2. If we consider the various isolated critical points in D (for which $\overline{D}$ is compact), there is clearly at most a finite number. Since D is the disjoint union of its connected components and D is locally connected, these components are themselves open. Thus, each isolated critical point lies in a connected open component of D. This yields the possibility of our assumption that the system has an isolated critical point which (as we did before) by translating the variables we can for convenience take to be the origin, O. In this way, the critical point, O, and solution, $X(t)$, both lie in D. Similarly, we can translate the t variable to have $O = X(0)$, i.e., $t_0 = 0$.

Definition 5.3.1. A critical point is called *stable* if for each ball B_R of radius $R > 0$ about O and in D, there exists a smaller ball B_r about O so that if $X(t_0)$ lies in B_r for some t_0, then $X(t)$ must lie in the larger ball for all $t \geq t_0$. Moreover, we shall say O is *asymptotically stable* if the path $X(t)$ approaches O as $t \to \infty$.

For a C^1 function E defined on D,

$$\frac{dE}{dt} = \sum_{i=1}^{n} \frac{\partial E}{\partial x_i} \frac{dx_i}{dt} = \sum_{i=1}^{n} \frac{\partial E}{\partial x_i} F_i = \langle \text{grad } E, X'(t) \rangle$$

Note that $X \mapsto \langle \text{grad } E, X'(t) \rangle$ is linear.

Let E be a C^1 function defined on D. We shall say E is *negative definite* if $E < 0$ everywhere on D except at (O), where $E(O) = 0$, E is called *negative semidefinite* if $E \leq 0$ everywhere on D except at (O), where $E(O) = 0$. Similarly, E is called *positive definite* (respectively positive semidefinite) if $-E$ is negative definite (respectively negative semidefinite). For example, a real-valued linear function on D is positive definite (respectively negative definite) if and only if all its eigenvalues are positive (respectively negative) and is semidefinite if and only if all its eigenvalues are greater than or equal to 0 (respectively lesser than or equal to 0).

Definition 5.3.2. A C^1 function E defined on D is called a Liapunov function if it is positive definite and if $\sum_{i=1}^{n} \frac{\partial E}{\partial x_i} F_i$ is negative semidefinite.

In applying this, the important point, of course, is to be able to find a Liapunov function.

Theorem 5.3.3. *Suppose there exists a Liapunov function E. Then the critical point O is stable. Moreover, if $\sum_{i=1}^{n} \frac{\partial E}{\partial x_i} F_i$ is actually negative definite, then as $t \to \infty$, $X(t)$ approaches O asymptotically.*

Proof.   Choose B_R small enough to lie in D. Since E is continuous and positive definite, it has a positive minimum value, m on B_R. Since E is continuous and $E(O) = 0$, there exists r so that $0 < r < R$ and $E < m$ on all of B_r. Suppose $X(t)$ intersects B_r at t_0. Then $E(t_0) < m$. Since $\frac{dE}{dt} \leq 0$, $E(t)$ is monotone decreasing, so $E(t) < m$ for all $t \geq t_0$. This means that $X(t)$ can never reach the boundary of B_R for $t \geq t_0$, i.e., lies in the open ball B_r.

Now, suppose $\sum_{i=1}^{n} \frac{\partial E}{\partial x_i} F_i$ is actually negative definite. Since E is positive definite, to prove asymptotic stability, all we need to do is show $E(t) \to 0$ as $t \to \infty$. Now, as before, $E(t)$ is monotone decreasing. Since E is positive definite, it is bounded from below by 0. This means that its greatest lower bound, $L = \lim_{t\to\infty} E(t)$. It remains only to show $L = 0$. Suppose $L > 0$. Choose a positive number $s < r$ so small that in B_s, $E < L$. Since $\sum_{i=1}^{n} \frac{\partial E}{\partial x_i} F_i$ is continuous and negative definite, it has a negative maximum value, say $-k$ in the closed region between the boundaries of B_r and B_s. This region contains the entire path $X(t)$ for $t \geq t_0$. By the fundamental theorem of calculus, $E(t) - E(t_0) = \int_{t_0}^{t} \frac{dE}{dt}$. Due to the inequality, $|\int f| \leq \int |f|$, this means that for all $t \geq t_0$, $E(t) \leq E(t_0) - k(t - t_0)$. Since the right side of this inequality tends to $-\infty$, so does the left which is impossible as $E \geq 0$. $\qquad\square$

In light of this, we now reconsider the pendulum and the predator–prey problems.

The ODE for the pendulum without damping and before making approximations was

$$\frac{d^2\theta}{dt^2} + \frac{g}{l}\sin(\theta) = 0.$$

Let $\nu = \frac{d\theta}{dt}$. Then $\frac{d\nu}{dt} = -\frac{g}{l}\sin(\theta)$. With damping, the second equation is $\nu = \frac{d\theta}{dt}$ and $\frac{d\nu}{dt} = -b\nu - \frac{g}{l}\sin(\theta)$, where $b > 0$, and so the 2×2 autonomous system is

$$\frac{d\theta}{dt} = \nu$$

and

$$\frac{d\nu}{dt} = -b\nu - \frac{g}{l}\sin(\theta).$$

Therefore, $(0,0)$ is a critical point and there are no others since $\theta \neq \pm\pi$. We take for the Liapunov function the *total energy* (kinetic and potential):

$$E(\theta, \nu) = \frac{1}{2}\nu^2 - \frac{g}{l}\cos(\theta).$$

Then

$$\frac{dE}{dt} = \nu\nu' + \frac{g}{l}\sin(\theta)\theta' = -b\nu^2,$$

which is negative definite since $b > 0$ (there is damping). Thus, Liapunov's theorem applies, so, as one would expect, $X(t)$

approaches $(0,0)$ asymptotically as $t \to \infty$. If there were no damping, then $b = 0$, and we would be in the negative semidefinite situation where $\frac{dE}{dt} = 0$, and so of course E would be constant.

We now turn to the predator–prey system. Here, everything takes place in the open first quadrant, D, and as we noted, the critical points were $(0,0)$ and $(\frac{c}{d}, \frac{a}{b})$. We take the latter for our O and take $E(x,y) = dx - c \log(x) + by - a \log(y)$ for our E.

Where is grad $E = 0$? That is, $d - c/x = 0$ and $b - a/y = 0$. Thus, $x = c/d$ and $y = a/b$, and $H_E(x,y) = \frac{c}{x^2} + \frac{a}{y^2}$, which is positive definite. Thus, $(\frac{c}{d}, \frac{a}{b})$ is an absolute minimum for E on D. Taking for our Liapunov function $dx - c \log(x) + by - a \log(y) - E(O)$, we can apply Liapunov's theorem. Due to the minimality of $E(O)$, one can show that the orbit through any point (other than O) in the open first quadrant is a simple closed curve. In Chapter 8, we will prove a general result to this effect which doesn't even need the details of a Liapunov function, but rather relies on the Poincare–Bendixson theorem (see Corollary 8.1.6). This implies that given any initial *positive* populations of predator and prey, these populations must oscillate cyclically. Neither will ever die out, nor will either grow indefinitely.

Chapter 6

Minding's Theorem, Sturm's Comparison Theorem and Other Results Concerning the Application of ODE to the Differential Geometry of Surfaces

6.1 The Exponential Map of a Surface, Geodesic Polar Coordinates and Minding's Theorem

Let $X(u,v)$ be a local surface in $\mathbb{R}^3$. That is to say, X is a smooth function on the parameter plane coordinatized by (u,v) taking values in $\mathbb{R}^3$. As u and v vary, say, over some open connected set in the parameter plane, the function X will trace out a surface S. For example, if $X(u,v) = Au + Bv + C$, where A, B and C are fixed vectors, then the surface is an arbitrary plane. Classical notation for the players in this are $\langle X_u, X_u \rangle = E$, $\langle X_u, X_v \rangle = F$ and $\langle X_v, X_v \rangle = G$, from which we can form the natural 2×2 symmetric matrix whose diagonal terms are E and G and whose off-diagonal term is F, thus giving a smooth symmetric matrix value function on the parameter space. In addition to smoothness of $X(u,v)$, we will also require that this matrix is positive definite. It goes by the name of *the first fundamental form* and tells us how to measure distance locally on the curved surface S. Then global distance is obtained by the integration of the local distances.

Here, we will be interested in local properties with special features, so we will be trying to find a well-chosen local coordinate system about a particular point p_0, namely, there is a fixed neighborhood of p_0 in the parameter plane in which each p in this neighborhood satisfies $(X_u(p), X_u(p)) = 1$ and $(X_u(p), X_v(p)) = 0$. Since $\det(g_{i,j}) > 0$, this forces $(X_v(p), X_v(p)) > 0$ and so taking square roots, we see $G(u, v) = g^2(u, v)$ for some smooth function g. When we have such a neighborhood, we call it a *local surface*. We will sometimes abuse the notation by using the same symbol for a set S in the parameter plane and its image on the surface.

In such a coordinate system, the length on the surface dominates the Euclidean length, i.e.,

$$\int_0^l \sqrt{u'(t)^2 + g^2(u, v)v'(t)^2}\, dt > \int_0^l u'(t)\, dt = u(l) - u(0). \qquad (6.1)$$

A curve in such a local surface is called a geodesic if it has the *shortest* length of all curves in S joining p_0 to another fixed (but nearby) point $p \in S$. If this holds for all $p \in S$, we call such a neighborhood a *geodesic* coordinate system. We may have to take a sub-neighborhood of S about p_0 to get everything we want from our neighborhood.

How can we produce such a local coordinate system? By the local existence theorem of ODE for each direction $v \in T_{p_0}$, the tangent plane to S at p_0, there is a local geodesic $\gamma(t, v)$ through p_0 so that $\gamma'(0) = v$ (see Appendix B and also Chapter 11 for the ODE). Then reparametrize this curve by its arc length, s.

Lemma 6.1.1. *Suppose the geodesic $\gamma(s, v)$ is defined for all $|s| < \epsilon$ and $c > 0 \in \mathbb{R}$. Then the geodesic $\gamma(s, cv)$ is defined for all $|s| < \frac{\epsilon}{c}$ and $\gamma(s, cv) = \gamma(cs, v)$.*

Proof. Let α be defined on $(-\frac{\epsilon}{c}, \frac{\epsilon}{c})$ by $\alpha(s) = \gamma(cs)$. Then $\alpha(0) = \gamma(0)$ and $\alpha'(0) = c\gamma'(0) = c^2 v$. By local existence and uniqueness (see Appendix B), α is a geodesic with initial conditions, $\gamma(0)$ and $c\gamma'(0)$. Hence, $\alpha(s) = \gamma(s, cv) = \gamma(cs, v)$.   □

If $v \neq 0 \in T_{p_0}$ has small enough norm, then $\gamma(||v||, \frac{v}{||v||}) = \gamma(1, v)$ is defined. We then define exp, the exponential map of a surface, by $\exp_{p_0}(v) = \gamma(1, v)$. We also take $\exp_{p_0}(0) = p_0$, our base point.

Since $v(a) = 0 = v(b)$, this last term is $-u(b)v'(b) + u(a)v'(a)$ and since $B \leq A$, and both are continuous, then unless $B = A$, $\int_a^b (B(t) - A(t))uv\,dt < 0$. Therefore, $u(a)v'(a) - u(b)v'(b) < 0$. On the other hand, since $v(b) = 0$ and v is positive just to the left of b, $v'(b) < 0$. Similarly, $v(a) = 0$ and v is positive just to the right of b, so $v'(a) > 0$. Therefore, since $u(a)$ and $u(b)$ are each ≥ 0, $u(a)v'(a) - u(b)v'(b) \geq 0$, a contradiction. $\qquad\square$

Taking $A = B$ we get the Sturm Comparison Theorem.

Corollary 6.2.2. *Suppose $u(t)$ and $v(t)$ are linearly independent solutions to $u''(t) + A(t)u(t) = 0$. Then between any two successive zeros $a < b$ of v, there is a zero of u.*

Proof. We need only observe that in the argument above, if $u(t)$ and $v(t)$ are linearly independent solutions, then $u(a)v'(a) - u(b)v'(b) \neq 0$, whereas the integral above is zero. This is because

$$u(a)v'(a) - u(b)v'(b) = \det \begin{bmatrix} u(a) & v'(a) \\ u(b) & v'(b) \end{bmatrix}.$$

$\qquad\square$

We now get two important geometric consequences of the Sturm comparison theorem.

Corollary 6.2.3. *Let S be a connected smooth surface in $\mathbb{R}^3$ whose curvature $K \geq c^2 > 0$. Then for any two points p and $q \in S$, $d(p,q) \leq \pi c$. In particular, if S is complete, it is compact.*

Proof. Let K be the curvature along a geodesic joining p and q, and consider the ODE $u'' + Ku = 0$ with initial conditions as above. Compare this to the solution v of the corresponding ODE $v'' + cv = 0$ (where $c \leq K(t)$ everywhere on the curve), whose solution is $a \sin ct + b \cos ct$. Since $v(0) = 0$, $b = 0$ and since $v'(0) = \pm 1$, $a = \pm 1/c$. Thus, $v(t) = \pm 1/c \sin(ct)$. The first positive zero of this function occurs at $t = \pi/c$. Therefore, the first zero of u is larger than π/c. But $\pi/\sqrt{K} < \pi/c$, so there are no zeros of u in $(0, \pi/\sqrt{K})$. $\qquad\square$

From this, we also get Hadamard's theorem in 3 space. It is why hyperbolic space cannot have conjugate points.

Proof.   Let $z = \frac{dy}{dx}$ be a new variable and consider the system of two first-order equations,

$$\frac{dy}{dx} = z,$$

$$y(x_0) = y_0,$$

and

$$\frac{dz}{dx} = -p(x)z - q(x)y + r(x),$$

$$z(x_0) = y_0'.$$

Thus, a solution is the same as a solution to this linear system and the initial conditions are also the same. So, it would be sufficient to prove the theorem for the linear system. Therefore, all we need to do is observe that the Picard theorem works for vector-valued functions (i.e., systems of ODEs), and when I is compact, it works *globally*. This is done in Appendix B. $\qquad\square$

In particular, we now know if x_0 is any point in I and $y(x_0)$ and $y'(x_0)$ are any numbers whatsoever, then there exists a unique solution $y(x)$ to ODE 7.0.1 with these initial conditions. We shall show that since there are precisely two independent conditions $\dim(V) = 2$. Let us write the system of linear equations in coordinate form:

$$\frac{dx}{dt} = a_{1,1}x(t) + a_{1,2}y(t),$$

$$\frac{dy}{dt} = a_{2,1}x(t) + a_{2,2}y(t),$$

with initial conditions $x(t_0) = x_0$ and $y(t_0) = y_0$, where $a_{i,j}$ are coordinates of a fixed matrix A and the vector-valued functions, $\mathbf{x}_1(t) = \begin{pmatrix} x_1(t) \\ y_1(t) \end{pmatrix}$ and $\mathbf{x}_2(t) = \begin{pmatrix} x_2(t) \\ y_2(t) \end{pmatrix}$, are a basis for $V = \mathbb{R}^2$ so that any solution

$$\mathbf{x}(t) = \begin{pmatrix} x(t) \\ y(t) \end{pmatrix}$$

in V can be *uniquely* written as

$$\mathbf{x} = c_1\mathbf{x}_1 + c_2\mathbf{x}_2.$$

In this way, for each point p_0, we have a well-defined function $\exp_{p_0}$ defined in a neighborhood V of 0 in T_{p_0}. By the fundamental theorem of ODE, $\exp_{p_0}$ is smooth since γ is (see Appendix B). Indeed, $\exp_{p_0}$ is actually a diffeomorphism from V to its image in S. To see this, we compute its derivative. For each $v \in V$,

$$\frac{d}{ds} \exp_{p_0}(sv)|_{s=0} = \frac{d}{ds} \gamma(s, v)|_{s=0} = v.$$

This means that $d(\exp_{p_0})|_0 = I$, the identity. Since this linear operator is non-singular, the inverse function theorem [10] tells us $\exp_{p_0}$ is a diffeomorphism of some sub-neighborhood of V containing the zero vector, which we again call V. On V, we have polar coordinates and since $\exp_{p_0}$ is a diffeomorphism, its inverse image gives polar coordinates on $U = \exp^{-1}(V)$. Here, U is a neighborhood of p_0 within S and the radial curves emanating from p_0 are geodesics. We call U a normal neighborhood of p_0 in S. Its coordinates called *geodesic polar coordinates* are (ρ, θ). The curves $\theta = $ a constant are geodesics.[1]

It is important to note that the exponential map of such a surface is related to, but different from, the usual power series exponential map (which we will be discussing in Chapter 9).

Proposition 6.1.2. *Let $E(\rho, \theta)$, $F(\rho, \theta)$ and $G(\rho, \theta)$ be the first fundamental form in these coordinates. Then $E = 1$, $F = 0$, $G > 0$. Moreover, $\lim_{\rho \to 0} G = 0$ and $\lim_{\rho \to 0} \sqrt{G}_\rho = 1$.*

In particular, since $F = 0$ this means that the curves ρ is a constant and intersect the curves θ is a constant orthogonally. This fact is called *Gauss' lemma*.

Proof. By definition of the exponential map, ρ is the arc length of the curve θ is constant. This means that $E = 1$. We now apply the compatibility relations. Since θ being constant is a geodesic, $0 = \frac{1}{2}E_\rho$. By the second of these relations, it follows that $F_\rho = 0$, and therefore, F depends only on θ. For each $p \in U$ (other than p_0), let $\rho = \phi(\theta)$ be its radial part, where $0 \le \theta < 2\pi$. Since these are radial geodesics

---

[1] However, note that to have a coordinate system, we require $\rho > 0$ and $0 < \theta < 2\pi$. Therefore, we have to remove the line $\theta = 0$. But this is harmless because removing this from U and its image from V still leaves a diffeomorphism.

$\gamma(s)$ passing through p (s being arc length),

$$F(\rho, \theta) = \left(\frac{d\phi}{d\theta}, \frac{d\gamma}{ds} \right).$$

Now, although $F(\rho, \theta)$ is not defined at $\rho = 0$ (namely, at p_0), if we fix θ, since $\phi(\theta)$ is constant, its derivative $\frac{d\phi}{d\theta} = 0$ and so the right side of the above equation is zero. Hence,

$$\lim_{\rho \to 0} F(\rho, \theta) = 0.$$

Since F is independent of ρ, this means that $F = 0$. Therefore, $EG - F^2 = G$ and, as we observed earlier, the first fundamental form is positive definite; therefore, $G = \det(g_{i,j}) > 0$.

Finally, under a smooth change of variables with derivative J, areas transform as follows: $\sqrt{EG - F^2} = \sqrt{E^*G^* - (F^*)^2}|\det J|$. When

$$u^* = \rho \cos(\theta), \quad v^* = \rho \sin(\theta),$$

then $|\det J| = \rho$. Hence, $\sqrt{G} = \rho\sqrt{E^*G^* - (F^*)^2}$. But in the $*$ coordinates, we are in *Euclidean* space, so $E^* = 1 = G^*$ and $F^* = 0$. Thus, $\sqrt{G} = \rho$.   □

From this, we get the following proposition.

Proposition 6.1.3. *Given $p_0 \in S$, there exists a neighborhood U about p_0 with the property that if $p \in U$, the unique geodesic joining p_0 to p has the shortest length of any curve in S joining p_0 and p.*

Let $K(p)$ denote the curvature at point p. We note that even though p_0 is technically excluded from the chart, K is certainly defined there.

Theorem 6.1.4. *Taking geodesic coordinates, we have $K(\rho, \theta) = -\sqrt{G}_{\rho,\rho}/\sqrt{G}$ (recall $G > 0$ everywhere), i.e., $\sqrt{G}$ satisfies the ODE,*

$$\sqrt{G(\rho,\theta)}_{\rho,\rho} + K\sqrt{G(\rho,\theta)} = 0.$$

In particular, if K is constant, then $\sqrt{G(\rho, \theta)}$ satisfies the second-order ODE in ρ, with constant coefficients. This will be proved in the course of proving Minding's theorem and, as the reader will see, we can draw some interesting consequences from it. First, we need to calculate the third-order Taylor series (see Appendix D) of $g(\rho, \theta) = \sqrt{G(\rho, \theta)}$ as a function of ρ.

Proposition 6.1.5. *Let K_{p_0} be the value of K at the origin. Then,*

$$g(\rho, \theta) = \rho - \frac{K_{p_0}}{3!}\rho^3 + \epsilon(\rho, \theta),$$

where $\lim_{\rho \to 0} \frac{\epsilon}{\rho^3} = 0$ uniformly in θ.

Proof. Since $g(\rho, \theta)_{\rho, \rho} = -K(\rho, \theta)g(\rho, \theta)$, differentiating with respect to ρ, we get

$$\frac{\partial^3 g}{\partial \rho^3} = -Kg_\rho - gK_\rho.$$

But as $\rho \to 0$, $g \to 0$. Hence,

$$\lim_{\rho \to 0} \frac{\partial^3 g}{\partial \rho^3} = -Kg_\rho.$$

Since $\rho \to 0$, the limit of g_ρ is 1:

$$\lim_{\rho \to 0} \frac{\partial^3 g}{\partial \rho^3} = -K(p_0). \tag{6.2}$$

Now, the Taylor expansion (see Appendix D) in ρ of the smooth function $g(\rho, \theta)$ is

$$g(\rho, \theta) = g(0, \theta) + \rho g_\rho(0, \theta) + \rho^2/2! g_{\rho, \rho}(0, \theta) + \rho^3/3! g_{\rho, \rho, \rho} + \epsilon(\rho, \theta).$$

Since the limiting value of each of the first and third terms is zero (see Taylor series in Appendix D), this proves the result. □

We can now calculate the arc length l of a geodesic circle of radius r. Since the points of discontinuity in geodesic polar coordinates is a set of area zero,

$$l = \int_0^{2\pi} g(r, \theta)d(\theta).$$

Using Proposition 6.1.5, we get, as a limiting value,

$$l = 2\pi r - \frac{\pi}{3} r^3 K(p_0). \tag{6.3}$$

Since curvature is actually an intrinsic quantity, this equation gives an intrinsic way of understanding the effect of the curvature at a point in terms of arc length of nearby geodesic circles. The arc length in the tangent space T_{p_0} of such a circle being $2\pi r$ and so here we get a correction term. If $K(p_0) > 0$, it makes l smaller, while if $K(p_0) < 0$, it makes l larger.

Solving for $K(p_0)$ yields $K(p_0) = \lim_{r \to 0} \frac{3}{\pi r^3}(2\pi r - l)$ which gives a way of calculating the curvature $K(p_0)$ from the arc length of small geodesic circles and proves K is intrinsic, i.e., depends only on the first fundamental form (and its derivatives) and not the embedding in $\mathbb{R}^3$.

We can also calculate the area A of a geodesic circle of radius r. In general, this is $\int \sqrt{EG - F^2} du dv$ which in our case tells us that $A = \int_0^r \int_0^{2\pi} g(\rho, \theta) d(\rho) d(\theta)$. Again, using the Taylor expansion, we get, for small r,

$$A = \pi r^2 - \frac{\pi}{12} r^4 K(p_0). \tag{6.4}$$

This gives another intrinsic way of understanding the effect of curvature at a point in terms of the area of small geodesic circles, since the area in the tangent space, T_{p_0}, is πr^2. So, here, again we get a correction. If $K(p_0) > 0$, it makes A smaller than the Euclidean area. If $K(p_0) < 0$, it makes A larger. Of course, if $K(p_0) = 0$, there is no effect!

Just as above, solving for $K(p_0)$ yields

$$K(p_0) = \lim_{r \to 0} \frac{12}{\pi r^4}(\pi r^2 - A),$$

giving a way of calculating the curvature from the areas of small geodesic circles.

Combining equations (6.3) and (6.4) gives us the isoperimetric inequality *with the error term* $\pi^2 r^4 K(p_0)$ (here, we can absorb the r^6 term into ϵ). The reader should compare this to the isoperimetric inequality without error term, i.e., in Euclidean space (see, e.g., [10]).

Corollary 6.1.6. $4\pi A - l^2 = \pi^2 r^4 K(p_0) + \epsilon$, *where* $\frac{\epsilon}{r^4} \to 0$ *as* $r \to 0$.

The reader should definitely compare this to the isoperimetric inequality without error term, i.e., in Euclidean space (see, e.g., [10]). We conclude this section with Minding's theorem on local isometries which is connected to classical second-order differential equations with constant coefficients. In effect, it gives a classification of surfaces of *constant* Gaussian curvature up to local isometry with the cases of positive, zero or negative curvature being the salient characteristic.

Theorem 6.1.7. *Any two regular surfaces S_1 and S_2 with the same constant Gaussian curvature are locally isometric. That is, there exist neighborhoods U_1 of p_1 in S_1 and U_2 of p_2 in S_2, which have the same first fundamental form. Conversely, if there exist such neighborhoods, then $K_1|U_1 = K_2|U_2$ since by definition isometries preserve the first fundamental form and therefore also the curvature.*

We emphasize that it is crucial here that the curvatures of both surfaces be constant. One cannot hope do better in the sense of getting a global isometry unless things are normalized by assuming that the surfaces are both simply connected.

Proof. As noted above, if K is constant and we take local geodesic coordinates, then $g(\rho, \theta)$ satisfies the second-order ODE in ρ, with constant coefficients,

$$g_{\rho,\rho} + Kg = 0.$$

We will separate the cases. First, suppose $K = 0$. Then $g_{\rho,\rho} = 0$, so g_ρ depends on θ alone. Thus, $g_\rho = f(\theta)$. On the other hand, since $g_\rho \to 1$ as $\rho \to 0$, this means that g_ρ is identically 1. Therefore, $g(\rho, \theta) = \rho + h(\theta)$ for some function $h(\theta)$. But then since $g \to 0$ as $\rho \to 0$, $h(\theta) = 0$ and $g = \rho$. Thus, the first fundamental form is $E = 1$, $F = 0$, $g = \rho$.

If $K > 0$, then we have the harmonic oscillator ODE whose solution we know well:

$$g(\rho, \theta) = a(\theta) \cos(\sqrt{K}\rho) + b(\theta) \sin(\sqrt{K}\rho).$$

Since $g \to 0$ as $\rho \to 0$, $a(\theta) = 0$ and $g_\rho \to 1$ as $\rho \to 0$, $b(\theta) = \frac{1}{\sqrt{K}}$. Thus, $E = 1$, $F = 0$ and $g = \frac{1}{\sqrt{K}} \sin(\sqrt{K}\rho)$. Similarly, if $K < 0$, then $E = 1$, $F = 0$ and $g = \frac{1}{\sqrt{-K}} \sinh(\sqrt{-K}\rho)$.

Now, let V_{p_0} and V_{q_0} be normal neighborhoods of p_0 and q_0, respectively. Let ϕ be a linear isometry from T_{p_0} to T_{q_0} and let $\psi : V_{p_0} \to V_{q_0}$ be defined by $\psi = \exp_{q_0} \circ \phi \circ \exp_{p_0}^{-1}$. Then ψ is a diffeomorphism and an isometry from V_{p_0} onto V_{q_0}.   □

Examples of Minding's theorem (and counterexamples to a possible global extension) are the cone and the cylinder as they are each locally isometric with the plane. This local isometry is implemented by unrolling each of them isometrically to create a planar surface. Such surfaces are called *developable*. The reason the isometry is only local is that each of these surfaces must be cut in order to unroll them.

6.2   Sturm's Comparison Theorem

We now turn to Sturm's comparison theorem which deals with ODEs of the second order, but now with *variable* coefficients and applications to surfaces of negative curvature (see [8]).

Consider the differential equation

$$\frac{d^2u}{dt^2} + A(t)u(t) = 0, \tag{6.5}$$

together with initial condition $u(0) = 0$ and $u'(0) = \pm 1$, where $A(t)$ is a smooth function of t on $[0, \infty)$. So, here, we have a second-order ODE, but now with variable coefficients. Due to these initial conditions, such a u is non-trivial, i.e., not identically zero.

Before getting to Sturm's comparison theorem, we make some clarifying remarks:

1. If u is a solution of this initial value problem, then $-u$ is also a solution of this initial value problem.
2. On a fixed closed subinterval $[a, b]$ of $[0, \infty)$, there can be at most finitely many, zeros of such a u. For if there were infinitely many, then there would be a sequence t_i with $u(t_i) = 0$, and by compactness, there would be a convergent subsequence (which we again call t_i converging to a point $t \in [a, b]$). Since u'' exists, u is C^1. In particular, u is continuous. Therefore, $u(t_i) \to u(t)$. Let $\epsilon > 0$. Then $|u(t) - u(t_i) + 0(t_i - t)| < \epsilon$. So, $u'(t) = 0$. Thus, u is trivial,

a contradiction. In particular, this shows that the zeros of this initial value problem are isolated.

3. Suppose u is a solution to equation (6.5) and $u(t) > 0$ for all $t \in [0, \infty)$. If $\int_0^\infty u(t)dt = \infty$, then u has infinitely many zeros on $[0, \infty)$. For if u had only finitely many such zeros, then at some point t_0 that is large enough, $u(t) > 0$ for all $t > t_0$. For $t \geq t_0$, let $v(t) = -\frac{u'(t)}{u(t)}$. Then $v'(t) = -\frac{uu'' - (u')^2}{u^2} = A(t) + v^2(t)$. Integrating this means that $v(t) - v(t_0) = \int_{t_0}^t A(t)dt + \int_{t_0}^t v^2(t)dt$. Since the second term is positive and the first goes to $+\infty$ if t is large enough, we see v goes to $+\infty$ for large t. Since v is positive, u and u' have opposite signs when t is large, but since v goes to infinity, this means that not only is u' negative in this range, but u is decreasing at an increasingly faster rate as $t \to \infty$. But then $\int_0^\infty u(t)dt$ would be finite, a contradiction.

Theorem 6.2.1. *Suppose $u(t)$ and $v(t)$ are, respectively, solutions to $u''(t) + A(t)u(t) = 0$ and $v''(t) + B(t)v(t) = 0$, with initial data as above and $A(t) \geq B(t)$ for all t. Then between any two successive zeros $a < b$ of v, there is a zero of u.*

Proof. Since $v(a) = 0 = v(b)$ and $v(t) \neq 0$ for $t \in (a, b)$, then as v is continuous, $v(t) > 0$ or $v(t) < 0$ everywhere on (a, b). We can eliminate the latter case by replacing v by $-v$, which still satisfies the ODE with initial conditions as before and will have the same zeros as the old v. Thus, we can assume $v(a) = 0 = v(b)$ and $v(t) > 0$ everywhere on (a, b). If u has mixed values on (a, b), then it must have a zero and we would be done. Otherwise, u is all positive or all negative on (a, b). By replacing u by $-u$, we can assume that $u(t) > 0$ on (a, b), and therefore, $u(t) \geq 0$ on $[a, b]$. We now argue by contradiction.

Suppose $u(t)$ is never zero on (a, b). Let $W(t) = u(t)v'(t) - v(t)u'(t)$ for $t \in [a, b]$. Then $W'(t) = (uv'' + u'v') - (vu'' + v'u') = uv'' - vu''$. Therefore, integrating

$$\int_a^b W'(t)dt = \int_a^b (B(t) - A(t))uvdt = (vu' - uv')|_a^b$$

$$= [v(b)u'(b) - u(b)v'(b)] - [(v(a)u'(a) - u(a)v'(a)].$$

Corollary 6.2.4. *Let S be a connected smooth complete surface in $\mathbb{R}^3$ whose curvature K is everywhere less than or equal to 0. Then no geodesic in S has conjugate points. In particular, if S is simply connected, then it is topologically Euclidean space.*

Of course, here, when we speak of a topological space, we mean disregarding all its features other than its topology.

Proof. Consider a solution u to the ODE, $u'' + Ku = 0$, with initial conditions as above, and let $c \geq K(t)$ as t varies over the curve $u(t)$. Compare this to the solution v of the corresponding ODE $v'' - c^2 v = 0$, whose solution is $a \sinh ct + b \cosh ct$. Since $v(0) = 0$, $b = 0$ and $v'(0) \neq 0$, a is arbitrary $\neq 0$. Thus, $v(t) = a \sinh(ct)$. Suppose $u(t_0) = 0$, where $t_0 > 0$. Then there is a $t_1 \in (0, t_0)$ so that $u(t_1) = 0$. But u has no zeros except at $t = 0$. Therefore, v has no zeros either. Hence, each geodesic is defined for all values of t and has no conjugate points. Taking a point p using local coordinates at p and the exponential map from $T_p \to S$, simple connectivity of S then shows that this is a diffeomorphism. $\qquad\square$

Here, we only get a local diffeomorphism. For a complete proof, we would need to use properties of simple connectivity which are to be found in topology courses and are beyond the scope of this book.

6.3 Geodesics

This next to final section of Chapter 6 is here because it applies to the geometry of surfaces. However, it requires some knowledge of the calculus of variations and could be regarded as a warm up to that subject which is treated in detail in Chapter 11. For this reason, the reader might want to skip this section until then.

Let $X'(s) = X_u u'(s) + X_v v'(s)$ be a curve parametrized by arc length $0 \leq s \leq l$ on a surface S. We call a curve on S joining $X(0)$ and $X(b)$ of shortest length a *geodesic*. Its length is given by the following formula:

$$L(X) = \int_0^b \sqrt{\sum_{i,j=1}^2 g_{i,j} X_i'(s) X_j'(s)}\, ds,$$

where $(g_{i,j})$ is the metric on S. As we shall see in Chapter 11, such curves must satisfy the Euler–Lagrange equations.[2]

$$\Phi_u(s) - \frac{d}{ds}\Phi_{u'}(s) = 0 \quad \Phi_v(s) - \frac{d}{ds}\Phi_{v'}(s) = 0,$$

where $\Phi(s) = \sqrt{\sum_{i,j=1}^{2} g_{i,j}X_i'(s)X_j'(s)}$.

Generally, one considers a geodesic to be a curve parametrized by arc length satisfying the Euler–Lagrange equations rather than a curve of shortest length (which is among them!). However, the square root in this integrand is inconvenient. For this reason, we consider the *Energy* functional instead (here, we are speaking only of the kinetic energy).

$$E(X) = \int_0^l \frac{1}{2} \left\| X'(s) \right\|^2 ds.$$

That is, we will show that when seeking geodesics, we can replace Φ above by the Φ associated with the energy, namely,

$$\Phi = \frac{1}{2} \sum_{i,j=1}^{2} g_{i,j}X_i'(s)X_j'(s),$$

and use the Euler equations of this new Φ instead. We emphasize that arc length is always the parameter.

We first observe that if $\Psi(u, v, u', v')$ satisfies the Euler equations and $\Phi(u, v, u', v') = \frac{1}{2}\Psi^2 i(u, v, u', v')$, then so does Φ.

Proof. For $\frac{\partial \Phi}{\partial u} = \Psi\frac{\partial \Psi}{\partial u}$ and since arc length is the parameter,

$$\Psi(s) = \left\| X'(s) \right\| = 1.$$

---

[2]Leonhard Euler (1707–1783) was a Swiss mathematician, physicist, astronomer and a bushel basket of other things as well. Although poor at repartee, he was probably the greatest thinker of the 18th century. Blind for much of his adult life, his productivity was enormous, so even limiting ourselves to mathematics, it would be futile to try to list all his accomplishments. Joseph Lagrange (1736–1813) was a French-Italian mathematical physicist. His treatise on analytical mechanics offered the most comprehensive treatment of classical mechanics since Newton and formed a basis for the development of mathematical physics in the 19th century. On the recommendation of Euler, Lagrange succeeded him as the Director of Mathematics at the Prussian Academy of Sciences in Berlin, Prussia, where he stayed for over 20 years.

Therefore,

$$\frac{\partial \Phi}{\partial u} = \frac{\partial \Psi}{\partial u}.$$

Similarly, $\frac{\partial \Phi}{\partial u'} = \Psi \frac{\partial \Psi}{\partial u'}$. Hence, $\Phi_u(s) - \frac{d}{ds}\Phi_{u'}(s) = \Psi_u(s) - \frac{d}{ds}\Psi_{u'}(s)$. Similarly, $\Phi_v(s) - \frac{d}{ds}\Phi_{v'}(s) = \Psi_v(s) - \frac{d}{ds}\Psi_{v'}(s)$. □

Definition 6.3.1. Given an embedded surface $X : U \times V \to S \subseteq \mathbb{R}^3$, we let

$$\nu(p) = \frac{X_u(p) \times X_v(p)}{\|X_u(p) \times X_v(p)\|}.$$

Since $\|X_u(p) \times X_v(p)\| \neq 0$, for each $p \in S$, we get a normal vector to the surface for each p, i.e., $\nu(p)$ is perpendicular to $T_p(S)$.

We now connect the Euler–Lagrange equations for the energy to the geometry of S by observing

$$\Phi_u(s) - \frac{d}{ds}\Phi_{u'}(s) = 0 \quad \text{if and only if } (X''(s), X_u(s)) = 0$$

and

$$\Phi_v(s) - \frac{d}{ds}\Phi_{v'}(s) = 0 \quad \text{if and only if } (X''(s), X_v(s)) = 0.$$

Proof. Since $\Phi = \frac{1}{2}(X', X')$, $\Phi_u = (X_u', X')$, while

$$\frac{d}{ds}\Phi_{u'}(s) = \frac{d}{ds}(X_u'(s), X'(s)) = (X''(s), X_u'(s)) + (X'(s), X_u'(s))$$

and similarly for v, v'. □

As a result, we get the following.

Corollary 6.3.2. $X(s)$ *is a geodesic if and only if* $(X''(s), X_u(s)) = 0$ *and* $(X''(s), X_v(s)) = 0$, *i.e.,* $X''(s)$ *is always perpendicular to* $T_p(X(s))$ *at every point on the curve. Hence, we write* $X''(s) = k_\nu(s)\nu(s)$.

Definition 6.3.3. k_ν is called the normal curvature.

As remarked above, in Chapter 11, we shall find the geodesics of various surfaces of revolution including the 2 sphere. There we will also find all geodesics in the Poincare upper half plane, H^+. This calculation will be based only on the hyperbolic metric and not on an embedding of H^+ in $\mathbb{R}^3$. Indeed, there is a famous theorem of Hilbert which proves H^+ cannot be *isometrically* embedded in $\mathbb{R}^3$, but can be isometrically embedded in $\mathbb{R}^4$.

Exercise 6.3.4. Let $(r\cos(\theta) \cdot r\sin(\theta), v)$ coordinatize a cylinder of radius $r > 0$, where $0 \le \theta < 2\pi$ and $v \in \mathbb{R}$. Choose points $p_0 = (r, 0, 0)$ and $q_0 = (r\cos(\theta) \cdot r\sin(\theta), v)$, where $v > 0$ and $0 < \theta < \pi$. Show that there are infinitely many geodesics joining p_0 and q_0, but only one of *minimal length*. Do the same for the cone.

6.4   A consequence of the Law of Cosines

We conclude this chapter with what we hope the reader will find an interesting consequence of the law of cosines due to Alexandrov.[3]

Theorem 6.4.1. *Let ABC be any triangle with sides a, b, c, each side being opposite the corresponding upper case angle and let M be the midpoint of side c so that $d(A, M) = \frac{c}{2} = d(M, B)$. Then,*

$$d^2(C, M) = \frac{1}{2}(d^2(C, A) + d^2(C, B)) - \frac{1}{4}d^2(A, B).$$

Proof.   Letting $u = d(C, M)$, we first rewrite the equation to be proved as $2u^2 + \frac{1}{2}c^2 = a^2 + b^2$. We apply the law of cosines to each of the smaller triangles ACM and BCM, where the angle C is divided as $C = \theta_A + \theta_B$, the first of these being the angle on the a side of C

---

[3]Pavel Alexandrov (1896–1982) was a Russian mathematician who made numerous important contributions to geometry and topology.

and the other the angle on the b side. Then

$$(c/2)^2 = a^2 + u^2 - 2au\cos(\theta_A)$$

and

$$(c/2)^2 = b^2 + u^2 - 2bu\cos(\theta_B).$$

Adding these, we get

$$c^2/2 = a^2 + b^2 + 2u^2 - 2u(a\cos(\theta_A) + b\cos(\theta_B)).$$

Now, draw perpendicular lines from A and from B to CM, extending CM where necessary. Letting F_A and F_B denote the foot of the respective perpendiculars, we get two right triangles AMF_A and BMF_B. Since angels AMF_A and BMF_B are vertical angles, these triangles are similar and since they have equal hypotenuses, they are congruent. In particular,

$$d(M, F_A) = d(M, F_B).$$

This means that $a\cos(\theta_A) + b\cos(\theta_B) = 2u$. Thus, $2u^2 + \frac{1}{2}c^2 = a^2 + b^2$, completing the proof. $\square$

Finally, we remark that the corresponding inequality,

$$d^2(C, M) \leq \frac{1}{2}(d^2(C, A) + d^2(C, B)) - \frac{1}{4}d^2(A, B)$$

(implying that this "triangle" is *thinner* than the one of the theorem and the law of cosines is actually an inequality, called the Alexandrov inequality, and plays a role in hyperbolic geometry).

Chapter 7

2×2 Linear Systems

We begin with the case of constant coefficients. Later, we will deal with variable coefficients.

We first consider the following second-order linear homogeneous ODE on some interval, $I = [a, b]$,

$$y'' + p(x)y' + q(x) = 0,$$

where p and q are continuous. Evidently, the solution set, V, is a real vector space (check!). Thus, any $\mathbb{R}$ linear combination of solutions is again a solution. We want to show its dimension is 2. That is, there are linearly independent solutions, y_1 and y_2, which generate V.

We now consider the more general associated non-homogeneous ODE

$$y'' + p(x)y' + q(x) = r(x),$$

where p, q, and r are continuous on $I = [a, b]$ and we want to prove the following result.

Theorem 7.0.1. *If x_0 is any point in I and $y(x_0)$ and $y'(x_0)$ are any numbers whatsoever, then there exists a unique solution $y(x)$ to this ODE with these initial conditions.*

Then everywhere on I the determinant of

$$\begin{pmatrix} x_1(t) & x_2(t) \\ y_1(t) & y_2(t) \end{pmatrix}$$

is $\neq 0$.

Denote this matrix by $B(t)$. Its determinant, $\det(B(t))$, is called the *Wronskian*[1] and is denoted $W(t)$. Since the Wronskian $W(t)$ is $\neq 0$, by Liouville's theorem, W satisfies its own ODE,

$$\frac{dW}{dt} = \operatorname{tr}(B(t))W(t),$$

which we already solved in Chapter 1. Hence, $W(t) = ce^{t\operatorname{tr}(B(t))}$, where c is a positive constant which confirms $W(t)$ is not zero on I. By Liouville's theorem the general solution of the original equation exists and V has dimension 2.

Obviously, these remarks can be extended to $n \times n$ systems and nth-order homogeneous, linear ODEs. Moreover, just as above, one could also deal with nth-order non-homogeneous, linear ODEs. For ODEs of order ≤ 2, we have already done this earlier in Chapter 2.

Here, we mention Liouville's theorem whose proof will be given in Chapter 9.

Consider the *matrix differential equation*

$$\frac{dX(t)}{dt} = A(t) \cdot X(t),$$

where $X(t)$ is the unknown matrix-valued function of t and $A(t) = (a_{ij}(t))$ is the matrix of coefficients. Here, t lies in some connected interval I containing 0 in $\mathbb{R}$ which could be a finite, or half infinite interval, or $\mathbb{R}$ itself. As initial condition, we take $X(0) = I$. We consider the linear homogeneous ODE, where $X(t)$ is the unknown path in the space of $n \times n$ real (or actually even complex) matrices and $A(t)$ is a given smooth $n \times n$ matrix-valued function of $t \in I$. In coordinates, this is a system of n^2 ODEs in n^2 unknowns, whose coefficients are $a_{ij}(t)$ and so always has local solutions.

---

[1] Josef Wronski (1776–1853) was a Polish mathematician who worked in ODE.

Liouville's theorem states that if $A(t)$ is real analytic and $X(t)$ satisfies the differential equation above, then $\det(X(t))$ satisfies its own (numerical) differential equation.

Theorem 7.0.2. *Let $A(t)$ be a real analytic function of t and suppose $X(t)$ satisfies the matrix ODE with initial condition as above. Then also on I,*

$$\frac{d\det(X(t))}{dt} = \operatorname{tr}(A(t))\det(X(t)),$$

with initial condition $\det(X(0)) = 1$.

This numerical ODE can be solved by separation of variables. It is closely connected to properties of the matrix exponential function which is dealt with in great detail in Chapter 9.

Exercise 7.0.3. As we have just seen, every second-order linear homogeneous (respectively non-homogeneous) ODE gives rise to an equivalent system of a pair of first-order linear homogeneous (respectively non-homogeneous) ODEs. Here, by equivalent, we mean the solutions of one coincide with the solutions of the other. Prove the converse, namely, given a pair of first-order linear homogeneous (respectively non-homogeneous) ODEs, there exists an equivalent second-order linear homogeneous (respectively non-homogeneous) ODEs.

Returning to the 2×2 case, let the pair of first-order linear homogeneous equations be determined by the 2×2-order matrix,

$$A = \begin{bmatrix} a & b \\ c & d \end{bmatrix}$$

and let λ be an eigenvalue of A, i.e.,

$$\lambda^2 - \operatorname{tr}(A)\lambda + \det(A) = 0.$$

Evidently, $\det(A) \neq 0$ if and only if zero is not an eigenvalue of A in which case A is invertible and the dimension of the solution space is 2. A natural question is: What are the possibilities for eigenvalues of A? Since the coefficients of the characteristic equation are real, the roots are either both real or are complex conjugates of one another and therefore distinct. In the real case, they could be distinct or coincide. Thus, we have all together three possible situations to consider:

I The roots are real and distinct.

II The roots are real and coincide. Here, there are two subcases:

II_1 A is conjugate over $\mathbb{R}$ to the triangular matrix,

$$\begin{bmatrix} \lambda & 0 \\ 1 & \lambda \end{bmatrix}.$$

II_1 A is conjugate over $\mathbb{R}$ to the scalar matrix,

$$\begin{bmatrix} \lambda & 0 \\ 0 & \lambda \end{bmatrix}.$$

III The roots are complex conjugates.

We also need to keep in mind that in Case I or II, some eigenvalue might be zero, and therefore, $\det(A) = 0$, which would affect the dimension of the solution space. However, as we shall see in case III, this cannot happen.

In Case I, A is diagonalizable over $\mathbb{R}$. Letting λ and μ be the two distinct real eigenvalues of A, then $(x, y) = (A_1 e^{\lambda t}, B_1 e^{\mu t})$ and $(x, y) = (A_2 e^{\lambda t}, B_2 e^{\mu t})$ are linearly independent fundamental solutions to the linear system. Therefore, here the general solution is

$$x = c_1 A_1 e^{\lambda t} + c_2 B_1 e^{\mu t},$$

$$y = c_1 A_2 e^{\lambda t} + c_2 B_2 e^{\mu t},$$

where c_1 and c_2 are arbitrary real constants. If neither of the eigenvalues is zero, then det $\neq 0$, so that $A_1 B_2 \neq A_2 B_1$, and since A_1, A_2, B_1, B_2 are all different from zero, $A_1/B_1 \neq A_2/B_2$. However, if one of the eigenvalues is 0, then the dimension of the solution space is less than 2.

A good example of Case I is the equation of the harmonic oscillator with damping given by

$$m\frac{d^2 x}{dt^2} + b\frac{dx}{dt} + kx = 0,$$

where $b > 0$ is the viscosity and $k > 0$ is the spring constant. The associated linear system is

$$\frac{dx}{dt} = y,$$

$$\frac{dy}{dt} = -k/mx + -b/my.$$

Here, the eigenvalues are 1 and $-b/m$ which are real and distinct since one is positive and the other negative.

In Case II_1, A is conjugate over $\mathbb{R}$ to

$$\begin{bmatrix} \lambda & 0 \\ 1 & \lambda \end{bmatrix},$$

i.e.,

$$\frac{dx}{dt} = \lambda x,$$

$$\frac{dy}{dt} = x + \lambda y.$$

Therefore,

$$\frac{dy}{dx} = x + \lambda y / \lambda x.$$

Writing this in differential forms,

$$\lambda(xdy - ydx) = xdx,$$

i.e., $\lambda d(\frac{y}{x})x^2 = xdx$, or $\lambda d(\frac{y}{x}) = \frac{dx}{x}$. Integrating yields

$$y = \frac{x \log x}{\lambda} + c,$$

where c is a constant.

Case II_2 is just a subcase of Case I.

In case III, A has complex conjugate eigenvalues $a \pm bi$. Since these roots are distinct, A is diagonalizable, but now over $\mathbb{C}$. It follows that $\det(A) = (a + bi)(a - bi) = a^2 + b^2 > 0$ since $b \neq 0$. Hence, A is invertible. Let B be the following real matrix:

$$\begin{bmatrix} a & b \\ -b & a \end{bmatrix}.$$

One easily checks that $a \pm bi$ are also the eigenvalues of B. Therefore, A and B are complex conjugates of one another and we can replace A by B. Thus, Case III is formally the same as Case I except that here the coefficients of the solutions are *complex*. Since $\lambda = a + ib$ and $\mu = a - bi$, here the two fundamental solutions are

$$(x, y) = \left(A_1^* e^{(a+ib)t}, B_1^* e^{(a-ib)t} \right)$$

and

$$(x,y) = \left(C_1^* e^{(a+bi)t}, D_1^* e^{(a-bi)t} \right),$$

where $A_1^*, B_1^*, C_1^*, D_1^*$ are in $\mathbb{C}$, and so we have to reality the situation. In dealing with the first, we have $A_1^* = A_1 + iA_2$, $B_1^* = B_1 + iB_2$. Then (7.0.1) says

$$(x,y) = ((A_1 + iA_2)e^{at}(\cos(bt) + i\sin(bt)),$$
$$(B_1 + iB_2)e^{at}(\cos(bt) - i\sin(bt))).$$

Thus,

$$x = (A_1 + iA_2)e^{at}(\cos(bt) + i\sin(bt))$$
$$= e^{at}([A_1\cos(bt) - A_2\sin(bt)] + i[A_2\cos(bt) + A_1\sin(bt)]),$$

and similarly,

$$y = e^{at}([B_1(\cos(bt) - B_2\sin(bt)] + i[B_2\cos(bt) + B_1\sin(bt)]).$$

Let $X_1 = e^{at}[A_1(\cos(bt) - A_2\sin(bt)]$ and $Y_1 = e^{at}[B_1(\cos(bt) - B_2\sin(bt)]$ be the real parts of x and y, respectively, where (x,y) is the first fundamental solution. Similarly, if (x,y) is the second fundamental solution, let $X_2 = e^{at}[A_2\cos(bt) + B_2\sin(bt)]$ and $Y_2 = e^{at}[B_2\cos(bt) + B_1\sin(bt)]$. Then both (X_1,Y_1) and (X_2,Y_2) are pairs of real functions and so are candidates to be fundamental solutions to our original real system of two real linear homogeneous ODEs. As these are the respective real parts of the solutions of the associated complex system of ODEs, their real parts are solutions of the original system!! Moreover, because of $V_\mathbb{C}$, a two-dimensional *complex* vector space, the real linear span of the two real parts, (X_1,Y_1) and (X_2,Y_2), is linearly independent over $\mathbb{R}$ and so span a real space of dimension 2. Thus, in Case III, letting c_1 and c_2 be arbitrary real constants, the general solution to the original system of ODEs is

$$x = e^{at}(c_1 A_1(\cos(bt)) - A_2(\sin(bt)) + c_2(A_2(\cos(bt) + A_1(\sin(bt))),$$
$$y = e^{at}(c_1 B_1(\cos(bt)) - B_2(\sin(bt)) + c_2(B_1\cos(bt) + B_2\sin(bt))).$$

An important special case of III is when $a = 0$, i.e., the matrix is

$$\begin{bmatrix} 0 & b \\ -b & 0 \end{bmatrix}.$$

Then

$$x = \big(c_1 A_1(\cos(bt)) - A_2(\sin(bt)) + c_2(A_2(\cos(bt)) + A_1(\sin(bt)))\big),$$
$$y = \big(c_1 B_1(\cos(bt)) - B_2(\sin(bt)) + c_2(B_1 \cos(bt) + B_2 \sin(bt)))\big).$$

This is the one case in 2×2 linear systems where the orbits in the x, y plane are closed periodic curves.

Remark 7.0.4. In the 2×2 case, we indicate why, even if the Wronskian is identically zero on an interval, I, the C^1 functions $y_1(x)$ and $y_2(x)$ may not be globally linearly dependent as functions on I.

To see this we first prove the following.

Proposition 7.0.5. *Suppose $y_1(x)$ and $y_2(x)$ are C^1 functions on I and Y is the matrix*

$$\begin{bmatrix} y_1(x) & y_2(x) \\ y_i'(x) & y_2'(x) \end{bmatrix}.$$

Then for each $x \in I$, $\det Y = y_1 y_2' - y_1' y_2 = W$, which we assume to be 0. From this, we will see that each $x \in I$ is contained in a proper subinterval, I_1, of I and I_1 itself contains a proper subinterval I_2, where y_1 and y_2 are linearly dependent. Thus, these functions will be locally *linearly dependent, but they may not be globally linearly dependent (as functions defined over all of I).*

Proof.   Let I_1 be any proper subinterval of I. If y_1 is identically zero on I_1, then clearly y_1 and y_2 are already linearly dependent on I_1. Otherwise, there must be some subinterval I_1 and a point x_0 in it where $y_1(x_0) \neq 0$. By continuity, there is a proper subinterval, I_2, of I_1 so that $y_1 \neq 0$ everywhere on I_2. Then on I_2, by the quotient rule,

$$\frac{d}{dx}\left(\frac{y_2}{y_1}\right) = \frac{y_1 y_2' - y_2 y_1'}{y_1^2} = 0.$$

Thus, on I_2, $\frac{y_2}{y_1}$ is a constant which is a dependence relation.   □

The proof of this proposition shows that to each subinterval I_1 of I, where $y_1 \neq 0$ on I_1, there corresponds a constant c so that $y_1 = cy_2$ on all of some proper subinterval, I_2. However, since c depends on I_2, we certainly can't hope to get a c that works everywhere on I.

This concludes our analysis of 2 × 2 *linear* (homogeneous) systems.

7.1   2 × 2 Systems with Variable Coefficients and the Qualitative Solutions of Second-Order ODEs

We now turn to more abstract 2 × 2 systems than those in the previous section and define some terminology. Here, we enter the more modern part of ODE where we may not be able to get exact solutions, but will have to be content with describing their salient features qualitatively or perhaps getting a visual on a computer screen.

Consider the system of ODEs

$$\frac{dx}{dt} = F(x, y), \tag{7.1}$$

$$\frac{dy}{dt} = G(x, y), \tag{7.2}$$

where F and G are C^1 functions on an open domain D in the x, y plane. We call the system *autonomous* because these functions are *independent* of t. We shall refer to $\mathbb{R}^2 = \{x, y\}$ as the *phase plane*. A *critical point* of (7.1) is a point (x_0, y_0) in the phase plane where

$$F(x_0, y_0) = 0 = G(x_0, y_0).$$

When we have linear equations, evidently $(0, 0)$ is always a critical point and if it's Jacobean,

$$J = \det\left(\begin{bmatrix} F_x & F_y \\ G_x & G_y \end{bmatrix}\right) \neq 0,$$

then this is the sole critical point (because the linearized equations are homogeneous). But in general, there could be others. (Here, the reader should recall the predator–prey equations of Volterra, where $\det(A) = 0$ gets us another critical point!)

This system of ODEs tells us the geometric fact concerning the vector field $\mathbf{v}(\mathbf{x}, \mathbf{y}) = \mathbf{i}F(\mathbf{x}, \mathbf{y}) + \mathbf{j}G(\mathbf{x}, \mathbf{y})$ in the domain D within the phase plane that, if a curve $(x(t), y(t))$ is a solution to the ODE (called an *integral curve*), then the tangent vector at each point has components $(F(x, y), G(x, y))$, or in physical terms, the velocity vector $\mathbf{v}(\mathbf{x}, \mathbf{y})$ at each point is $(F(x, y), G(x, y))$. The *critical points* of the system are those $(x, y) \in D$ where $\mathbf{v}(\mathbf{x}, \mathbf{y}) = \mathbf{0}$, i.e., where the velocity is zero.

We now come to the type of consideration which led Poincare to invent topology! Suppose C is any smooth closed curve in the phase plane (not necessarily simple) on which no critical point lies. For each point on C, $\mathbf{v} = \mathbf{F}(\mathbf{x}, \mathbf{y})\mathbf{i} + \mathbf{G}(\mathbf{x}, \mathbf{y})\mathbf{j}$ is a non-zero vector and therefore has a definite polar angle, θ. As a point varies counterclockwise over C and finally returns to its original position, θ varies by $2\pi n$ where $n \in \mathbb{Z}$. The integer n is called the *index* of C and is independent of the initial choice of the point on the curve. Suppose C shrinks continuously to a smaller *simple* closed curve within C *without passing through any critical point* (this shrinking is called a *homotopy*). Since its index varies continuously and the index takes its values in the *discrete* set, $2\pi\mathbb{Z}$, it cannot change because a continuous map from a connected set to a discrete set is constant. Thus, it remains n. We now apply this idea to arrive at the following simple but quite important fact.

Theorem 7.1.1. *Given a continuous vector field $\mathbf{v}(\mathbf{x}, \mathbf{y})$ and a closed disk D both in the plane, if the index of the boundary curve $S = \partial(D)$ is non-zero, then there must be a critical point for $\mathbf{v}$ inside D.*

Proof. Suppose there are no critical points inside D. Then S can be continuously deformed anywhere inside D without passing through any critical points. In particular, we can continuously deform S to be arbitrarily near the center 0 of D. The index of this small curve about 0 is zero. Since the index remains constant under continuous deformations, it must have been zero to begin with, a contradiction. $\square$

To illustrate the power of this simple idea, we use it to prove the fundamental theorem of algebra. First, a few remarks. There is no harm in considering a monic polynomial since we assume the polynomial has degree $n \geq 1$, so the leading coefficient is non-zero. Dividing by it, we get a monic polynomial whose roots are the same as before. Moreover, once we know there is a root, we can factor the polynomial and get one of one lower degree and apply induction to see the polynomial factors completely into linear polynomials. An important and easily verifiable fact about the polynomial $p(z) = z^n + a_1 z^{n-1} + \cdots + a_n$, which we leave to the reader, is that $\lim_{z \to \infty} \frac{|z|^n}{p(z)} = 1$.

Corollary 7.1.2. *The complex polynomial $z^n + a_1 z^{n-1} + \cdots + a_n$ must have a complex root.*

Proof. Let $\mathbf{v}(\mathbf{z}) = \mathbf{z^n} + \mathbf{a_1 z^{n-1}} + \cdots + \mathbf{a_n}$. This is a continuous vector field on the plane whose zeros are the roots of the polynomial we are interested in. $\mathbf{v_t}(\mathbf{z}) = \mathbf{z^n} + \mathbf{t}(\mathbf{a_1 z^{n-1}} + \cdots + \mathbf{a_n})$, where $0 \le t \le 1$ is a continuous deformation from $\mathbf{v}$ to $\mathbf{v_0}$, i.e., from $p(z)$ to z^n. The fact that $\lim_{z \to \infty} \frac{|z|^n}{\mathbf{v(z)}} = 1$ tells us that if r is large enough, the index of our vector field is the index of the vector field associated with z^n, which is clearly $n \ne 0$. By Theorem 7.1.1, there is an r sufficiently large so that D_r has a zero on $\mathbf{v}$. $\qquad\square$

Given an autonomous system, as before, if $F \ne 0$ on D, to find explicit solutions, it is sometimes helpful to eliminate t completely by dividing by F (see, e.g., Volterra's predator–prey equations as well as the way we handled the case of multiple roots just above). Thus,

$$\frac{dy}{dx} = \frac{G(x,y)}{F(x,y)}$$

is a first-order ODE which says that the slope of the tangent line at every point on an integral curve is $\frac{G(x,y)}{F(x,y)}$.

We now give an example of a nonlinear autonomous system which will serve as a guide though the heuristics of our analysis which follows:

$$\frac{dx}{dt} = -y + x(1 - x^2 - y^2),$$

$$\frac{dy}{dt} = x + y(1 - x^2 - y^2).$$

To better understand this system, we first introduce *polar coordinates* $x = r\cos(\theta)$, $y = r\sin(\theta)$. Now, differentiate the inverse relations $r^2 = x^2 + y^2$ and $\tan(\theta) = \frac{y}{x}$ when $x \neq 0$. This yields $r\frac{dr}{dt} = x\frac{dx}{dt} + y\frac{dy}{dt}$ and

$$\sec^2(\theta)\frac{d\theta}{dt} = r^2/x^2\frac{d\theta}{dt} = x\frac{dy}{dt} - y\frac{dx}{dt}.$$

Now, multiply the first equation of (7.1) by x and add it to the second equation of (7.1) and also multiply the second equation by y and add it to the first. After the smoke clears, by some miracle, the variables separate:

$$\frac{dr}{dt} = r(1 - r^2)\frac{d\theta}{dt} = 1,$$

and each gives an *exact* integral!! Thus, solving these equations individually tells us that either $r = 0$, or $\theta = t + t_0$ and $r = (1 + ce^{-2t})^{-1/2}$, where t_0 and c are constants. Of course, we may as well take $t_0 = 0$ and start t at 0. What do these equations tell us? $r = 0$ is just a point and this is not a path we are interested in. Suppose $c = 0$. Then $r \equiv 1$ and the solution is a closed orbit, namely, it traces out the unit circle going counterclockwise starting from 1. If $c > 0$, then $r > 1$ and $r \to 1$ in a spiral outside of the circle and spiraling toward the circle as $t \to \infty$. If $c < 0$, then $r < 1$ and $r \to 1$ in a spiral inside the circle and spiraling toward the circle as $t \to \infty$.

All this suggests that with *appropriate hypotheses*, the general situation might be that aside from single points (which would be considered degenerate solutions), what one has are bounded solutions (like the circle) or solutions which spiral into such a closed curve from either the inside or the outside. If we normalize $t_0 = 0$, then the *global* solution is *unique*.[2] We now get more specific. We are interested in the following global question: Does an integral curve such as (7.1) have a *simple closed* path? This is clearly equivalent to the question of whether or not (7.1) has a simple *periodic* solution.

By a closed curve, we mean a curve C in D so that $C : [0, l] \to \mathbb{R}$ and $C(0) = C(l)$, l being the arc length of the curve. Here, we

---

[2]It also raises the interesting topological question: Does a general *smooth simple closed* curve, C, in the plane always have an inside (the bounded part) and an outside (the unbounded part) so that $\mathbb{R}^2 \setminus C$ is the disjoint union of the two (connected sets)? The famous Jordan curve theorem says *yes*.

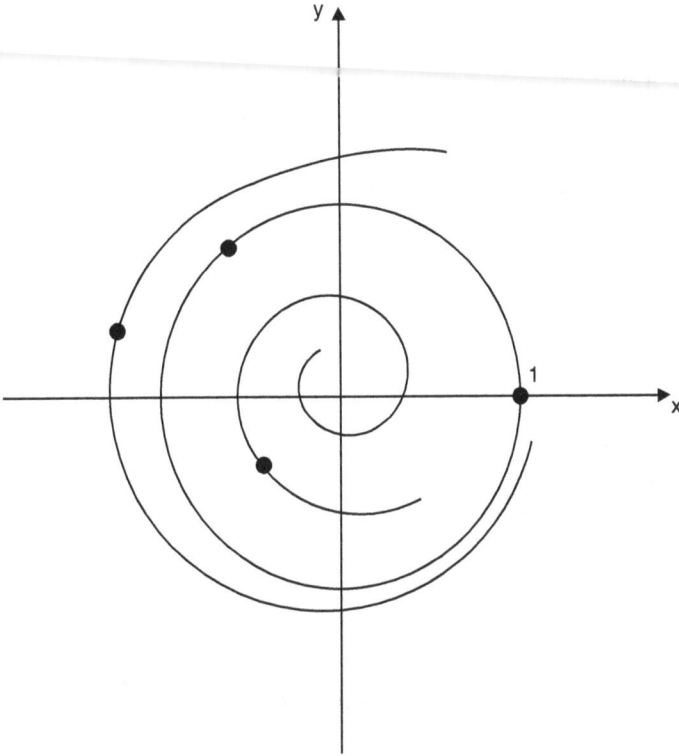

Figure 7.1. A periodic non closed integral curve (hence containing a spiral) with two spirals.

assume the curve is C^1. The curve is called *simple* if this function is injective, i.e., the curve doesn't cross over itself. We have already defined *periodicity* of a function. We remind the reader concerning this notion because now we are talking about a solution to an autonomous dynamical system (7.1).

Definition 7.1.3. A plane curve $(x(t), y(t))$ (non-constant and defined for all $t \in \mathbb{R}$) is called periodic if there exists a $\omega > 0$ such that

$$(x(t), y(t)) = (x(t + \omega), y(t + \omega))$$

for all t. The smallest such positive number ω is called the period of the function $(x(t), y(t))$. Since the curve is rectifiable, such a ω is its arc length l.

The Poincare–Bendixson theorem in the following chapter gives rather general sufficient conditions for a region D in the plane to contain a closed path. It was the Poincare–Bendixson theorem and similar issues which led Poincare to topology. Contrarywise, the following proposition due to Lienard, which requires the use of Green's theorem (see Appendix C), gives a criterion sufficient for the *non-existence* of a closed path as a solution.[3]

Proposition 7.1.4. *If $\frac{\partial F}{\partial x} + \frac{\partial G}{\partial y}$ is always ≥ 0, or always ≤ 0, then an integral curve of equation* (7.1) *cannot have a closed path unless $\frac{\partial F}{\partial x} + \frac{\partial G}{\partial y}$ is actually identically zero.*

Proof. Let C be an integral curve of the system which is a closed path in D, say of length l, and Ω be its interior together with C itself. Then $C = \partial(\Omega)$ (the boundary of Ω). Suppose, respectively, that $\frac{\partial F}{\partial x} + \frac{\partial G}{\partial y}$ were positive or negative at some point on C. Then it would be positive (respectively negative) in some interval U of Ω about that point. To avoid circumlocutions, let us consider the positive case. (The negative case is handled similarly). Then

$$0 < \iint_U \frac{\partial F}{\partial x} + \frac{\partial G}{\partial y} dxdy \leq \iint_\Omega \frac{\partial F}{\partial x} + \frac{\partial G}{\partial y} dxdy.$$

By Green's theorem,

$$\int_C (Fdy - Gdx) = \iint_\Omega \frac{\partial F}{\partial x} + \frac{\partial G}{\partial y} dxdy.$$

But along C, in differential forms, $dx = Fdt$ and $dy = Gdt$, so

$$\int_C (Fdy - Gdx) = \int_0^l (FGdt - GFdt).$$

Since multiplication is commutative, this is zero, a contradiction. $\square$

We now turn to Lienard's theorem itself, the proof of which is taken from [13], and which will occupy the remainder of this chapter. Before doing so, we shall need the following additional fact.

---

[3]George Green (1793–1841) was a self-taught physicist and mathematician as well as a miller by trade. His work greatly contributed to modern physics.

Lemma 7.1.5. *Let ϕ be a real-valued differentiable function defined on $\mathbb{R}$ and ϕ' be its derivative. Then ϕ is odd if and only if ϕ' is even and ϕ is even if and only if ϕ' is odd.*

Proof. Suppose ϕ is even. Then $\phi(x) = \phi(-x)$ for all x. Differentiating and using the chain rule, we see $\phi'(x) = -\phi'(-x)$ and similarly for the second statement. □

Lienard's theorem gives a qualitative description of solutions to the following family of second-order ODEs:

$$\frac{d^2 f}{dt^2} + f(x)\frac{df}{dt} + g(x) = 0,$$

where f and g are real-valued C^1 functions defined on $\mathbb{R}$, satisfying the following conditions:

1. g is an odd function and $g(x) > 0$ when $x > 0$.
2. f is an even function and the odd function, $F(x) = \int_0^x f$, has exactly one positive zero at $x = a$, is negative in $(0, a)$, and is positive and non-decreasing for $x > a$.
3. $F(x) \to \infty$ as $x \to \infty$.

Under these conditions, we will prove (7.1) has a unique solution which is a closed path in the phase plane and that this path is approached spirally by every other path, as $t \to \infty$. Thus, it is a special (and quite practical) case of the Poincare–Bendixson theorem of the following chapter. Of course, by Theorem 7.1.1, we know that inside this closed path, there must be a critical point.

Proof. Using the above lemma and the fundamental theorem of calculus, we see that F is an odd function. Let us write (7.1) as an autonomous 2×2 system,

$$\frac{dx}{dt} = y,$$

$$\frac{dy}{dt} = -g(x) - f(x)y.$$

By the fundamental existence theorem (see Appendix B), (7.1) has a unique solution in the phase plane. By condition 1, $g(0) = 0$ and

$g(x) \neq 0$ when $x \neq 0$, so the origin is the only critical point of (7.1). Further,

$$\frac{d^2 x}{dt^2} + f(x)\frac{dx}{dt} = \frac{d}{dt}\left[\frac{dx}{dt} + \int_0^x f\,d\right] = \frac{d}{dt}[y + F(x)].$$

Letting $z = y + F(x)$, we see that the system (7.1) is equivalent to the simpler

$$\frac{dx}{dt} = z - F(x),$$

$$\frac{dz}{dt} = -g(x)$$

in the (x, z) plane. Since F is continuous, the map $(x, y) \mapsto (x, y + F(x)) = (x, z)$ is a homeomorphism. In the new coordinates, g and F remain odd functions and $(0, 0) \mapsto (0, F(0)) = (0, 0)$. Since F is odd, none of the objectives of what we are trying to prove has been compromised by the change of coordinates. In particular, we only have to concern ourselves with the right half plane since points on a curve in the left hand plane can be located from the corresponding point in the right half plane by reflection through the origin.

Dividing in (7.1), we get

$$\frac{dz}{dx} = \frac{-g(x)}{z - F(x)}.$$

If a path becomes horizontal, i.e., $\frac{dz}{dx} = 0$, then $-g(x) = 0$, so $x = 0$. That is, at that point, the path must cross the z-axis. If a path becomes vertical, i.e., $z - F(x) = 0$, then $z = F(x)$, and the path must cross the curve $z = F(x)$. Inspection of the signs in (7.1) tells us that all curves are directed to the right above the curve $z = F(x)$ and are directed to the left below $z = F(x)$ and they move downward or upward according to whether $x > 0$ or $x < 0$. This means that the curve $z = F(x)$, the z-axis and the vertical line through any point Q on the right half of the curve $z = F(x)$ can only be crossed in the directions of the arrows indicated above.

Suppose a solution of (7.1) defining the path C through Q is chosen so that it corresponds to $t = 0$. Then as t increases into positive values, a point $(x(t), y(t))$ moves down and to the left until it crosses the z-axis at a point R, whereas if t decreases into negative values, the point $(x(t), y(t))$ rises to the left until it crosses the z-axis at a point P. Let b be the x coordinate of Q and denote the part of the path C from P to Q by C_{PQ} or C_b.

From the symmetry, it follows that when the path C_{PQ} continues beyond P and R into the left half plane, we will get a closed path if and only if P and R are equidistant from the origin. Hence, to show there is a unique closed path, it suffices to show there is a unique value of b making P and R equidistant from the origin. To do so, let $G(x) = \int_0^x g$ so that G is an even function and let $E(x, z) = \frac{1}{2}z^2 + G(x)$ so that $E(0, x) = G(x)$. Then along any path at all, we have

$$\frac{dE}{dt} = g(x)\frac{dx}{dt} + z\frac{dz}{dt},$$

and on our path, this is

$$-[z - F(x)]\frac{dz}{dt} + z\frac{dz}{dt} = F(x)\frac{dz}{dt}.$$

Thus, in differential forms, $dE = Fdz$.

We now compute the line integral, I_b, of Fdz along C_b from P to R:

$$\int_{PR} Fdz = \int_{PR} dE = E_R - E_P = \frac{1}{2}(d(OR)^2 - d(OP)^2).$$

Therefore, to complete the proof, it suffices to show that there is a unique b with $I_b = 0$. If $b \leq a$, then F and dz are both negative, so $I_b > 0$ and so C_b cannot be closed. Now, suppose $b > a$. We want to show here that I_b is a decreasing function of b. To do this, we split I_b as the sum of two parts, $I_b = I_1(b) + I_2(b)$, where $I_1(b) = \int_{PS} Fdz + \int_{TR} Fdz$ and $I_2(b) = \int_{ST} Fdz$. Since F and dz are both negative as C_b is traversed from P to S and from T to R, $I_1(b) > 0$, whereas when going from S to T, $F > 0$ and $dz < 0$, so $I_2(b) < 0$. By equation (7.1),

$$Fdz = F\frac{dz}{dx}dx = \frac{-g(x)}{z - F(x)}dx.$$

The effect of increasing b is to raise arc PS and lower arc TR which decreases the magnitude of the right side of the equation above for any $x \in (0, a)$. Since the limits of integration for $I_1(b)$ are fixed, this results in a decrease of $I_1(b)$ and because $F(x)$ is positive and non-decreasing to the right of a, an increase in b gives rise to an increase of the positive number $-I_2(b)$, i.e., a decrease in $I_2(b)$. Thus, $I_b = I_1(b) + I_2(b)$ is a decreasing function of b in the region, $b \geq a$. We will now show $I_2(b) \to -\infty$ as $b \to \infty$. If L in the figure above is fixed and K is to its right, then

$$I_2(b) = \int_{ST} F \, dz < \int_{NK} F \, dz < \int_{NK} F \, dz \leq - \int_{LM} + \int_{LN}.$$

Since $\int_{LN} \to \infty$ as $b \to \infty$, $I_2(b) \to \infty$ as $b \to \infty$. Hence, $I(b) = 0$ for a unique b_0 the so there is a unique $b = b_0$, for which C_{b_0} is closed.

Finally, if $b < b_0$, then $OR > OP$. So, because of the symmetry, the paths inside C_{b_0} spiral out toward C_{b_0}. Similarly, if $b > b_0$, then $OR < OP$, so the paths outside C_{b_0} spiral in toward C_{b_0}.   □

We now apply what we have just learned to the following example called the *van der Pol* equation:

$$\frac{d^2 f}{dt^2} + \mu(x^2 - 1)\frac{df}{dt} + x = 0,$$

where $\mu > 0$.

Corollary 7.1.6. *Equation (7.1) has as a solution, a unique closed periodic path in the x, y plane. This path is approached asymptotically by every other non-trivial solution.*

This is because here $F(x) = \mu\frac{x^3 - 3x}{3}$ and $g(x) = x$, and one checks that all conditions of Lienard's theorem are satisfied, where $a = \sqrt{3}$. Therefore, equation (7.1) has such a solution.

Here is one further example: Consider the second-order ODE

$$\frac{d^2 x}{dt^2} + a\frac{dx}{dt} + bx = f(x'),$$

where $a, b > 0$ and f is a smooth monotone increasing odd function asymptotic to parallel lines $y = c > 0$ and $y = -c$. Let $y = \frac{dx}{dt}$.

Then the equivalent 2 × 2 system is

$$\frac{dx}{dt} = y,$$

$$\frac{dy}{dt} = -ay - bx - f(y).$$

One checks that the only zero of this vector field is $(0,0)$. The derivative (matrix) of this dynamical system is

$$\begin{bmatrix} 0 & 1 \\ -b & -a - f'(0) \end{bmatrix}.$$

Its eigenvalues are $\frac{f'(0)-a}{2} \pm \sqrt{\frac{(f'(0)-a)^2}{4} - b}$. Thus, we require $f'(0) > a$ and $(f'(0) - a)^2 \geq 4b$ and c to be adjusted to f. For example, we could take $f = tan^{-1}$, $c = \pi/2$, $0 < a < 1$ and $b < \frac{(1-a)^2}{4}$. Then the solution will have an asymptotic limit cycle.

We remark that, in general, an example may have a band containing various closed integral curves as well as spirals which can be either inside or outside the band, or some of each.

Chapter 8

Autonomous Dynamical Systems and the Poincare–Bendixson Theorem

In this chapter, we shall begin our discussion in $\mathbb{R}^n$, but when we get down to business, of necessity, we shall restrict ourselves to $\mathbb{R}^2$. We consider autonomous systems

$$\frac{dx_i}{dt} = F_i(x_1, \ldots, x_n),$$

where F_i are C^1 real-valued functions on an open domain D in $\mathbb{R}^n$. This can be expressed more conveniently using vector notation

$$\frac{dX}{dt} = F(X),$$

where $F : \mathbb{R}^n \to \mathbb{R}^n$ is a smooth vector-valued function. As before, the term "autonomous" means that F is *independent* of t (i.e., only depends on the space coordinates). A solution is an $\mathbb{R}^n$-valued function $X(t)$ (i.e., a curve in $\mathbb{R}^n$) whose derivative with respect to t satisfies this equation, i.e., the ODE $X' = F(X)$. At least locally, for $t \in I$, where I is a neighborhood of 0 in $\mathbb{R}$, the existence and uniqueness of such solutions are addressed in Appendix B. An important part of this is the *smooth* dependence on initial conditions. This means that a solution, $\phi_t(X)$, is $C^1(I \times \mathbb{R}^n)$, i.e., $\frac{\partial \phi}{\partial X}$ and $\frac{\partial \phi}{\partial t}$ both exist and are continuous.

A smooth vector field in $\mathbb{R}^n$ (or on an open subset, U, of it) is a smooth map $U \to \mathbb{R}^n$. Thus, F of the ODE just above can also be regarded as a smooth vector field on $\mathbb{R}^n$. Solutions to such systems are curves in $\mathbb{R}^n$ where the tangent vector at every point, X, on the curve equals $F(X)$. These solution curves can be written ϕ_t in $\mathbb{R}^n$ where the tangent vector at every point X on the curve in $\mathbb{R}^n$ is $F(X)$. Thus, we have a smooth function $\phi : I \times \mathbb{R}^n \to \mathbb{R}^n$. If ϕ extends to a smooth function which we again call ϕ, then

$$\phi : \mathbb{R} \times \mathbb{R}^n \to \mathbb{R}^n.$$

We call this a *smooth autonomous dynamical system* if it satisfies the following conditions (i.e., these are the axioms):

1. $\phi(0, X) = X$ for all $X \in \mathbb{R}^2$.
2. For all $s, t \in \mathbb{R}$ and $X \in \mathbb{R}^2$, $\phi(t, \phi(s, X)) = \phi(t + s, X)$.

It will be convenient to write $\phi_t(X)$ rather than $\phi(t, X)$. Here, t will be thought of as time and ϕ as a dynamical system with X its initial condition. In this notation, solutions to such systems are curves ϕ_t in $\mathbb{R}^n$ where the tangent vector at every point X on the curve in $\mathbb{R}^n$ is $F(X)$ and ϕ is called a *flow*.

The reader should note that these axioms of Chapter 8 become theorems in Chapter 9 when for a fixed $\mathbf{v}$, we take $\phi_t(X)$ to be $\text{Exp}(tX)(\mathbf{v})$.

As usual, an *equilibrium point* of a smooth autonomous dynamical system is a point $Y \in \mathbb{R}^n$ where $F(Y) = 0$. Suppose Y is *not* an equilibrium point of our dynamical system. If there exists a positive real number τ such that $\phi_\tau(Y) = Y$, the smallest such (positive) τ is called the period of ϕ, and such a ϕ is called a *periodic solution*. Thus,

$$\phi_{t+\tau}(Y) = \phi_t(\phi_\tau(Y)) = \phi_t(Y),$$

so that the curve returns to its original position.

We now come to the concepts of *invariance* and ω *limits*.

Definition 8.0.1. A set $A \subset \mathbb{R}^n$ is called *positively invariant* if $\phi_t(A) \subset A$ for all $t > 0$. Similarly, A is called negatively invariant if $\phi_t(A) \subset A$ for all $t < 0$. If A is both positively and negatively invariant, we just say A is ϕ invariant (since $\phi_0 = I$).

Clearly, a periodic orbit is invariant.

Definition 8.0.2. Let $X \in \mathbb{R}^n$. The ω limit of $\phi_t(X)$ is all $Y \in \mathbb{R}^n$ for which there exists a strictly increasing sequence $t_n \uparrow \infty$ with

$$\lim_{t_n \uparrow \infty} \phi_{t_n}(X) = Y.$$

$\omega(X)$ denotes the set of all its ω limits. Similarly, $\alpha(X)$ denotes the set of all ω limits as $t_n \downarrow -\infty$.

If an orbit, A, is a *closed* set, then it contains all its limit points and therefore all its ω limits. Hence, $\alpha(X) = A = \omega(X)$ for every $X \in A$.

We now prove some useful lemmas concerning flows of C^1 autonomous systems in $\mathbb{R}^n$.

Lemma 8.0.3. *Suppose Y and Z lie on the same solution curve to $X' = F(X)$. Then $\omega(Y) = \omega(Z)$ and $\alpha(Y) = \alpha(Z)$.*

Proof. Let $W \in \omega(Y)$. Then there exists $t_n \uparrow \infty$ with $\lim_{t_n} \phi_{t_n}(Y) = W$. Since Z lies on the same solution curve as Y, there is a real s so that $\phi_s(Y) = Z$. Choose m so that $t_m > s$. Then for all $k \geq m$,

$$\phi_{t_k}(Y) = \phi_{t_k - s + s}(Y) = \phi_{t_k - s}\phi_s(Y) = \phi_{t_k - s}(Z).$$

So, $\lim_{k \uparrow \infty} \phi_{t_k}(Y) = \lim_{k \uparrow \infty} \phi_{t_k - s}(Z)$. Therefore, $\omega(Y) \subseteq \omega(Z)$. By symmetry $\omega(Z) \subseteq \omega(Y)$. A similar argument proves $\alpha(Y) = \alpha(Z)$.

$\square$

Lemma 8.0.4. *Suppose $X \in \mathbb{R}^n$ lies on the solution curve to $X' = F(X)$. Then $\omega(X)$ and $\alpha(X)$ are invariant sets.*

Proof. Let $Y \in \omega(X)$. Then there is $t_n \uparrow \infty$ such that $\lim_{t_n}(\phi_{t_n}(X) = Y$. Then for any $s \in \mathbb{R}$, choose n large enough so that $t_n > s$. Since ϕ is C^1 in both variables, we know that ϕ is continuous. Hence,

$$\lim_{n \uparrow \infty} \phi_{s+t_n}(X) = \lim_{n \uparrow \infty} \phi_s(\phi_{t_n})(X) = \phi_s \lim_{n \uparrow \infty} (\phi_{t_n})(X)$$

$$= \phi_s(Y) \in \omega(X).$$

Since $s + t_n \uparrow \infty$, $\phi_s(Y) \in \omega(X)$ and because s was arbitrary, $\omega(X)$ is invariant. The proof for $\alpha(X)$ is similar. $\square$

Lemma 8.0.5. *Suppose $X \in \mathbb{R}^n$ lies on a solution curve to $X' = F(X)$. Then $\omega(X)$ and $\alpha(X)$ are closed sets in $\mathbb{R}^n$.*

Proof.   Let Y be in the closure of $\omega(X)$ and $B(r, X)$ be the open ball of radius $r > 0$ about X. Then taking $r = \frac{1}{n}$, for every $n \geq 1$, $B(\frac{1}{n}, X) \cap \omega(X)$ is non-empty. That is, there exists a sequence $Y_n \in \omega(X)$ so that $d(Y_n, X) \leq \frac{1}{n}$. Since $Y_n \in \omega(X)$, for every m, there exists an arbitrary large $t_m \in \mathbb{R}$ so that $d(\phi_{t_m}(X), Y_m) < \frac{1}{m}$. Thus, we have a subsequence t_k of t_n so that $d(\phi_{t_k}(X), Y_k) < \frac{1}{k}$ and $t_{k+1} > t_k + 1$. Hence, for all k,

$$d(Y, \phi_{t_k}(X)) \leq d(Y, Y_k) + d(Y_k, \phi_{t_k}(X)) \leq \frac{2}{k}.$$

Therefore, we have a found a sequence $t_k \uparrow \infty$, whose limit is Y. Thus, $Y \in \omega(X)$, so $\omega(X)$ is closed. A similar proof holds for $\alpha(X)$.   □

Lemma 8.0.6. *Suppose A is a closed, positively invariant set and $X \in A$, then $\omega(X) \subseteq A$. Similarly, if B is a closed negatively invariant set and $Y \in B$, then $\alpha(Y) \subseteq B$. Finally, if C is a closed invariant set, then C contains the ω and α limits of each of its points.*

Proof.   Clearly, it suffices to prove the first statement, so suppose A is a closed, positively invariant set and $X \in A$. For $P \in \omega(X)$, there exists a sequence $t_n \uparrow \infty$ with $\phi_{t_n}(X) = P$. Since these t_n go to $+\infty$, they must be positive for $n \geq m$. Throw away this finite number of them and we can assume the t_n are all positive and since A is closed and positively invariant, all $\phi_{t_n} \in A$ and these converge to P, we see $P \in A$. Thus, $\omega(X) \subseteq A$.   □

Lemma 8.0.7. *The set $\omega(X)$ is compact if and only if there exists an $s \in \mathbb{R}$ so that $\{\phi_t(X) : t \geq s\}$ is bounded. The set $\alpha(X)$ is compact if and only if there exists an $s \in \mathbb{R}$ so that $\{\phi_t(X) : t \leq s\}$ is bounded.*

Proof.   As usual, we need to only prove the first statement. Suppose there exists an $s \in \mathbb{R}$ so that $\{\phi_t(X) : t \geq s\}$ is bounded; say, it is contained in the closed ball $B = B(r_0, \phi_s(X))$, where r_0 is some fixed positive number. As such, B is positively invariant for the point $\phi_s(X)$. By Lemma 8.0.6, $\omega(\phi_s(X)) \subseteq B$. Since $\phi_s(X)$ lies on the same solutions set as X, Lemma **??** tells us that $\omega(\phi_s(X)) = \omega(X)$ so that actually, $\omega(X)$ itself is contained in B and therefore is bounded. On the other hand, it is closed by Lemma 8.0.5. Therefore, by Heine–Borel, it is compact.

Conversely, suppose $\omega(X)$ is compact. Then of course it is bounded, so for any $Y \in \mathbb{R}^n$, there is a positive radius r so that the compact set $\omega(X) \subseteq B(r, Y)$. Taking for Y, $\omega(\phi_s(X))$, we see that for each real s, there exists an $r > 0$ so that $\omega(X) \subseteq B(r, \omega(\phi_s(X)))$.

Now, we want to show for each s, $\{\phi_t(X) : t \geq s\}$ is bounded. Suppose for some s, this was unbounded. Let $P \in \omega(X)$. Then there exists a sequence $t_n \uparrow \infty$ so that $\lim t_n(\phi_{t_n}(X)) = P$. Now, the fixed open bounded set $B(r, \omega(\phi_s(X)))$ contains some subsequence of this, which we also denote by $\phi_{t_n}(X)$, which is unbounded. Therefore, this must cross the boundary of B infinitely many times (since if it only crossed finitely many times, we could just modify the radius of B and get a bounded sequence contradicting our assumption). Consider this infinite sequence of boundary points. By compactness of the boundary, this sequence must have a convergent subsequence which converges to a point Q on the boundary. But $\phi_{t_n}(X) \subseteq B$, $Q \in \partial(B)$ and B is disjoint from $\partial(B)$, a contradiction. $\qquad\square$

The following lemma actually has nothing to do with dynamical systems, it's just topology.

Lemma 8.0.8. *Let U and V be disjoint, non-empty, open sets in $\mathbb{R}^n$. Then $\partial(U) \cap V = \emptyset$ and $\partial(V) \cap U = \emptyset$.*

Proof. Suppose, for example, there was a $P \in V$ and also $P \in \partial(U)$. Since V is open, there would be a neighborhood N of P contained in V. On the other hand, P lies on the boundary of U, so N intersects $\partial(U)$. Therefore, $V \cap \partial(U)$ is non-empty. Since every point on $\partial(U)$ can be approximated by points of U and V is open, this contradiction proves the lemma. $\qquad\square$

We are slowly coming to the point as we can now combine these items to get a theorem on smooth autonomous dynamical systems.

Theorem 8.0.9. *Let $X \in \mathbb{R}^n$. If $\omega(X)$ is compact, then $\omega(X)$ is non-empty and connected.*

Proof. Suppose $\omega(X)$ is compact for $X \in \mathbb{R}^n$. By Lemma E, there exists an $s \in \mathbb{R}$ so that $\{\phi_t(X) : t \geq s\}$ is bounded. Consider the increasing sequence of real numbers $a_n = s + n$ for positive integers n. Since $a_n > s$, Lemma E tells us that the sequence $\phi_{a_n}(X)$ is bounded. By Bolzano–Weierstrass, there is a subsequence $\phi_{b_n}(X)$ which converges to P that lies in $\omega(X)$, which is compact. Thus, $\omega(X)$ is non-empty.

Now, suppose contrariwise that $\omega(X)$ were not connected. Then $\omega(X) = U \cup V$, where U and V are open, non-empty and disjoint subsets of $\omega(X)$. Let $Y_u \in U$ and $Y_v \in V$. Then as before there exist increasing sequences s_n and t_n so that $\lim_{n \uparrow \infty} \phi_{s_n}(X) = Y_u$ and $\lim_{n \uparrow \infty} \phi_{t_n}(X) = Y_v$. Since U and V are open, we can choose n large enough so that $\phi_{s_n}(X_u) \in U$ and $\phi_{t_n}(X_v) \in V$. (Renumbering for convenience so that we are starting from $n = 1$.) Now, choose a sequence r_i so that the odd terms come from s_n and the even ones come from t_n. Since $\phi_{s_n}(X_u)$ and $\phi_{t_n}(X_v)$ each converge the subsequences, $\phi_{r_i}(X_u)$ and $\phi_{r_i}(X_v)$ also converge (to X_u and X_v, respectively) and since they are both subsequences of ϕ_{r_i} and we are in a Hausdorff space, $X_u = X_v$. Thus, $U \cap V$ is non-empty, a contradiction. $\qquad\qquad\square$

8.1   The Case $n = 2$

Now that we have established the general facts, we shall need concerning dynamical systems in n dimensions, we turn to $n = 2$ and the final and most general result of this section, namely, the Poincare–Bendixson theorem.[1] First, what is so special about $n = 2$? To explain, we recall the definition of a simple closed curve. This is a curve $X(t)$ in $\mathbb{R}^n$, which for convenience begins at $t = 0$ and where $X(0) = X(l)$ at *exactly one* value $l > 0$, i.e., no crossing over on itself. Now, in the plane, we have the Jordan curve theorem which is a fundamental fact about the topology of the plane. It tells us that after removing a continuous simple closed curve (in our case, a C^1 simple closed curve) from the plane, what is left has exactly two connected components, namely, the inside (the bounded part) and the outside (the unbounded part). This is false if $n \geq 3$ since

---

[1]Ivar Bendixson (1861–1935) was a Professor at Stockholm University from 1913 to 1927 and was later rector there. He worked on set theory and differential equations.

one can smoothly go from the "inside" to the "outside", i.e., there is no inside or outside. Continuous simple closed plane curves are called Jordan curves (of course, they need not be convex.) If γ is a Jordan curve, $Int(\gamma)$ and $Ext(\gamma)$ denote, respectively, the interior and exterior of γ.

We continue our study of autonomous dynamical systems, but now only in two dimensions. As we shall see, this will significantly limit the possibilities for $\omega(X)$ and $\alpha(X)$.

First, a construction: Let $X' = F(X)$ be a C^1 autonomous system in $\mathbb{R}^2$ and X_0 be a non-equilibrium point, i.e., $F(X_0) \neq 0$ and V be a unit vector based at X_0 which is perpendicular to $F(X_0)$. Now, define $h : \mathbb{R} \to \mathbb{R}^2$ by $h(u) = X_0 + uV$. This is a line in the plane passing through X_0 and of course perpendicular to $F(X_0)$ called the *transverse line* $l(X_0)$. It has been constructed so as not to be tangent to $X' = F(X)$ at X_0. Now, not being tangent to something is an open condition, so by continuity, there is a small ϵ neighborhood about X_0 so that all X with $d(X, X_0) < \epsilon$ are also not tangent to the curve at $F(X_0)$. Pulling back by the continuous F and then applying h to the inverse image gets us an interval about X_0 on the transverse line, all pointing perpendicular to V called a *local section* $\mathcal{S}$ at X_0. Of course, also, $F(X) \neq 0$ everywhere on $\mathcal{S}$.

We now give a description of the flow on the local section, $\mathcal{S}$, at X_0 by constructing a smooth map $\Psi : N \to \mathcal{S}$, where N is a neighborhood of O in $\mathbb{R}^2$. For $(s, u) \in N$, we define $\Psi(s, u) = \phi_s(h(u))$. Then Ψ maps the vertical line segment $(0, u) \in N$ to $\mathcal{S}$ and maps horizontal lines in N to pieces of solution curves of the system. $D\Psi$ takes the constant vector field $(1, 0)$ in N to the vector field $F(X)$. Choosing N sufficiently small ensures that Ψ is injective on N. In this way, using our section, $\mathcal{S}$, we have smoothly reparametrized a small neighborhood of any non-equilibrium point by N.

Proposition 8.1.1. *Let $\mathcal{S}$ be a local section at X_0 and suppose $\phi_{t_0}(Z_0) = X_0$. Then for any neighborhood W of Z_0, there is a smaller neighborhood U of Z_0 and a continuous function $\tau : U \to \mathbb{R}$ such that $\tau(Z_0) = t_0$ and $\phi_{\tau(X)} \in \mathcal{S}$ for all $X \in U$.*

Proof. Define the function $\eta : \mathbb{R}^2 \to \mathbb{R}$ by $\eta(Y) = \langle Y, F(X_0) \rangle$. We recall that if Y belongs to the transverse line $l(X_0)$, this means that $Y = X_0 + V$, where $\langle V, F(X_0) \rangle = 0$. Thus, $\langle Y, F(X_0) \rangle = \langle X_0, F(X_0) \rangle$. Now, define $H : \mathbb{R}^2 \times \mathbb{R} \to \mathbb{R}$ by $H(X, t) =$

$\eta(\phi_t(X)) = \langle \phi_t(X), F(X_0) \rangle$. Then since $\phi_{t_0}(Z_0) = X_0$, $H(Z_0, t_0) = \langle X_0, F(X_0) \rangle$. Moreover,

$$\frac{\partial H}{\partial t}(Z_0, t_0) = \|F(X_0)\| \neq 0.$$

Applying the implicit function theorem (see [10]) gets us a smooth function $\tau : U_1 \to \mathbb{R}$ defined on a neighborhood $U_1 \subseteq \mathbb{R}^2$ of (Z_0, t_0) such that $\tau(Z_0) = t_0$ and

$$H(X, \tau(X)) = H(Z_0, t_0) = \langle X_0, F(X_0) \rangle.$$

This means, $\phi_{\tau(X)}(X)$ lies on $l(X_0)$. Then choosing a sufficiently small neighborhood $U \subseteq U_1$ about Z_0, we get $\phi_{\tau(U)} \subseteq \mathcal{S}$.   □

Definition 8.1.2.

1. A finite or infinite sequence X_i in $\mathbb{R}^2$ is called *monotone* if there exists a non-negative increasing sequence t_i in $\mathbb{R}$ and for all i, $X_i = \phi_{t_i}(X_0)$.
2. Given a local section $\mathcal{S}$, we call finite or infinite sequence X_i in $\mathcal{S}$ monotone if for all i, X_i is between X_{i-1} and X_{i+1}.

Lemma 8.1.3. *Suppose that $X' = F(X)$ is a C^1 autonomous system in $\mathbb{R}^2$, $\mathcal{S}$ is a local section and Y_i is a sequence of points of $\mathcal{S}$ which all lie on the same solution curve $X(t)$. If Y_i are monotone on $X(t)$, then they are monotone along $\mathcal{S}$.*

Proof. If not, then there exists an i such that Y_{i+1} is between Y_{i-1} and Y_i. Let C be the segment of the curve $X(t)$ between Y_{i-1} and Y_i, and D be the line segment in $\mathcal{S}$ joining Y_{i-1} and Y_i. Then $C \cup D$ is a simple closed curve in $\mathbb{R}^2$. Moreover, for all $X \in D$, $F(X)$ points away from $\int(C \cup D)$. Since $\phi_t(Y_i)$ cannot intersect C and cannot enter the interior of $C \cup D$, it must remain in the exterior of $C \cup D$. On the other hand, because Y_i are monotone on $X(t)$, there exists a $t \in \mathbb{R}$ so that $\phi_t(Y_i) = Y_{i+1}$. This means Y_{i+1} is in the exterior of $C \cup D$ which contradicts the fact that Y_{i+1} is in D.   □

Lemma 8.1.4. *If $Y \in \omega(X)$ or $Y \in \alpha(X)$ for some $X \in \mathbb{R}^2$, then the flow, $\phi_t(Y)$, crosses a local section $\mathcal{S}$ at most once.*

Proof. As usual, we deal with the $\omega(X)$ case. So, $Y \in \omega(X)$ for some $X \in \mathbb{R}^2$. Since by Theorem 8.0.9, $\omega(X)$ is non-empty, it suffices to show there can't be two crossings. Suppose the flow, $\phi_t(Y)$, crosses a local section S at two distinct points Y and Z. Let U and V be respective disjoint local neighbourhoods of Y and Z. Then there exist increasing real sequences u_n and v_n so that $\lim_{n \to \infty} \phi_{u_n}(X) = Y$ and $\lim_{n \to \infty} \phi_{v_n}(X) = Z$. Eventually, these end up in U and V, respectively. This means there exist $u_{n_1} < v_{n_2} < u_{n_3}$ where $\phi_{u_{n_1}}(X)$ and $\phi_{u_{n_3}}(X)$ are in U and $\phi_{v_{n_2}}(X)$ is in V. This means that the sequence $\phi_{u_{n_1}}(X), \phi_{u_{n_3}}(X), \phi_{v_{n_2}}(X)$ is monotone along $X(t)$, contradicting 8.1.3. $\qquad\square$

Finally, we turn to the Poincare–Bendixson theorem itself. Since $n = 2$, here we return to the original notation for an autonomous dynamical system.

Theorem 8.1.5. *Let*

$$\frac{dx}{dt} = F(x, y),$$

$$\frac{dy}{dt} = G(x, y),$$

where F and G are C^1 functions on a region D with compact closure in the phase plane, $\mathbb{R}^2$ of the ODE. If Ω is the limit set of the ODE and contains no equilibrium points, then Ω is a closed orbit.

What this tells us is that if Ω contains no critical points of 8.1.5 and $C = (x(t), y(t))$ is a smooth integral curve of 8.1.5 in Ω, then for $t \geq 0$, either

1. C is itself a closed periodic path or
2. C spirals toward a such closed path in Ω as $t \to \infty$ (either from the inside or the outside). In particular, Ω always has a closed integral path. Another way to say 2 is, there are points on C which are Euclidean close, but whose distance along the solution curve is large and ever increasing.

Proof. Suppose for $X \in \Omega$, $\omega(X)$ is compact and $Y \in \omega(X)$. Then we shall show Y lies on a closed orbit γ and $C = \omega(X)$. Since $Y \in \omega(X)$ we know from Theorem 8.0.9 that $\omega(Y)$ is non-empty. By Lemma D, it is a subset of $\omega(X)$, so let $Z \in \omega(Y)$ and S be a local section at Z. By 8.1.1 and 8.1.4, the solution through Y meets S at exactly one point. On the other hand, there is a sequence

$\{t_n\} \to \infty$ such that $\phi_{t_n}(Y) \to Z$. There being infinitely many of these, there exist real numbers $r > s$ so that $\phi_r(Y)$, $\phi_s(Y) \in \mathcal{S}$. Therefore, $\phi_r(Y) = \phi_s(Y)$, and hence, $\phi_{r-s}(Y) = Y$, while $r - s > 0$. But because $\omega(X)$ contains no equilibrium points, Y must lie on a closed orbit.

We will now show that if C is a closed orbit in $\omega(X)$, then $C = \omega(X)$, and for that it is enough to show $\lim_{t\to\infty} d(\phi_t(X), C) = 0$, where d is the distance from $\phi_t(X)$ to the compact set C. Let $\mathcal{S}$ be a local section at $Y \in C$ and $\epsilon > 0$. Then there is a set $t_n, t_0 < t_1 < \cdots$ such that

1. $\phi_{t_n}(X) \in \mathcal{S}$,
2. $\phi_{t_n}(X) \to Y$,
3. $\phi_t(X)$ not in $\mathcal{S}$ for $t_n < t < t_{n+1}$ and $n \geq 1$.

By 8.1.3, $X_n = \phi_{t_n}(X)$ is a monotone sequence in $\mathcal{S}$ that converges to Y. We first show there is an upper bound for the set of positive numbers, $\{t_{n+1} - t_n\}$. For suppose $\phi_\tau(Y) = Y$, where $\tau > 0$, then for X_n sufficiently near Y, there is some t with $|t| \leq \epsilon$, so that $\phi_{\tau+t}(X_n) \in \mathcal{S}$. This forces $t_{n+1} - t_n \leq \tau + \epsilon$.

Let $\beta > 0$ be any (small) positive real number. By continuity of solutions with respect to initial conditions (see Appendix B), there exists a $\delta > 0$ such that if $\|Z - Y\| < \delta$ and $|t| \leq \tau + \epsilon$, then $\|\phi_t(Z) - \phi_t(Y)\| < \beta$. In particular, the distance from the solution $\phi_t(Z)$ to C is less than β for all t satisfying $|t| \leq \tau + \epsilon$. Let n_0 be large enough so that for all $n \geq n_0$, $\|X_n - Y\| < \delta$. Then $\|\phi_t(X_n) - \phi_t(Y)\| < \beta$ if $|t| \leq \tau + \epsilon$ and $n \geq n_0$. Now, let $t \geq t_{n_0}$ and $n \geq n_0$ satisfy $t_n \leq t \leq t_{n+1}$. Then since $|t - t_n| \leq \tau + \epsilon$,

$$d(\phi_t(X), C) \leq \|\phi_t(X) - \phi_{t-t_n}(Y)\| = \|\phi_{t-t_n}(X_n) - \phi_{t-t_n}(Y)\| < \beta.$$

This means that the distance from $\phi_t(X)$ to C is less than β if we take t large enough and completes the proof. $\qquad\square$

We remark that in the course of the proof of the Poincaré–Bendixson theorem, we have also proved that if C is an ω limit cycle, then there must exist an $X \in \Omega$, but not in C such that $\lim_{t\to\infty} d(\phi_t(X), C) = 0$. That is, the solution spirals toward C as $t \to \infty$. Let $\mathcal{S}$ be a local section at $Z \in C$. Then there is an interval $I \subseteq \mathcal{S}$ disjoint from C bounded by $\phi_{t_0}(X)$ and $\phi_{t_1}(X)$ with $t_0 < t_1$

and not meeting any solution through X for $t_0 < t < t_1$. This proves the following corollary.

Corollary 8.1.6. *Let D be a bounded open set in the phase plane, $\mathbb{R}^2$, and $\frac{dX}{dt} = F(X)$ be a C^1 autonomous dynamical system and C a solution. Then C cannot be a spiral and so by Poincare–Bendixson theorem, it must be a simple closed curve.*

Proof. Suppose C were a spiral. Then looking at the polar angles 0 and π, we see that at these angles, $\frac{dx}{dt} = 0$. Therefore, there exist t_ns either tending to ∞ or $-\infty$ where $x'(t_n) = 0$. From one of our earlier results, this means that $x'(t) = 0$ for all $t \geq t_1$ (or all $t \leq t_1$). Similarly, looking at points with polar angle $\frac{\pi}{2}$ (or $\frac{3\pi}{2}$), there exists t_2 so that $y'(t) = 0$ for all $t \geq t_2$. Therefore, for all $t \geq \max(t_1, t_2)$, both are zero, i.e., $X'(t) = 0$. This means that $X(t)$ is constant after $\max(t_1, t_2)$. This contradicts the assumption that $X(t)$ is a spiral. $\square$

Corollary 8.1.7. *Let C be a ω limit cycle. If $C = \omega(X)$ where X is not in C, then X has a neighborhood U such that $C = \omega(Y)$ for all $Y \in U$. That is, the set $\{Y : \omega(Y) = C\} \setminus \{C\}$ is open.*

Corollary 8.1.8. *Let C be a closed orbit and U be an open region in the interior of C. Then U either contains an equilibrium point or a limit cycle.*

Proof. This is because if D were a compact region, $D \subseteq U \cup C$. Since no solution in U can cross C if U contained no limit cycle and no equilibrium point, then for any $X \in U$, $\omega(X) = C = \alpha(X)$ by the Poincare–Bendixson theorem. If S is a local cross-section at $Z \in C$, there would be sequences $t_n \to \infty$ and $s_n \to -\infty$ such that $\phi_{t_n}(X)$ and $\phi_{s_n}(X)$ are both in S and both tend to Z as $s, t \to \infty$, contradicting Lemma D. $\square$

One can now sharpen Corollary 8.1.8. But note that we have actually already proved this in Chapter 7 and in a simpler way!

Corollary 8.1.9. *Let C be a closed orbit that forms the boundary of an open set U. Then U must contain an equilibrium point.*

Proof. Suppose U contained no equilibrium points. If there were only finitely many closed orbits in U, choose one that bounds the region of minimal area. There are no closed orbits in this region. This contradicts Corollary 8.1.8. Now, suppose there are infinitely many closed orbits in U. If $X_n \to X$ and each X_n lies in a closed orbit, then X itself must lie in a closed orbit, for otherwise the solution through X would spiral toward a limit cycle by the Poincare–Bendixson theorem since we are assuming there are no equilibrium points in U. By Corollary 8.1.7 so would the solution through a nearby X_n. Let $\nu \geq 0$ be the greatest lower bound of the areas of the regions enclosed by U and C_n be a sequence of closed orbits enclosing regions of area ν_n with $\lim_{n\to\infty} \nu_n = \nu$. For each n, choose $X_n \in C_n$. Since $C \cup U$ is compact by selecting a subsequence, we may assume $X_n \to X \in U$. Since there are no equilibrium points in U, by the Poincare–Bendixson theorem, X would lie in in a closed orbit C_0 bounding a region of area ν. As $n \to \infty$, C_n gets arbitrarily close to C_0. This means that the areas of the regions between C_n and C_0 tends to 0. The argument of the first paragraph of this proof takes care of the rest. $\square$

We conclude our discussion of autonomous systems with an application to topology itself. This is a version of the Brouwer fixed point theorem[2] when $n = 2$.

Theorem 8.1.10. *Let K be a compact convex set in $\mathbb{R}^2$ and f be a continuous map of K to itself. Then f has a fixed point.*

Proof. First, we look for fixed points on the boundary, $\partial(K)$. If there is one, we are done. Otherwise, each point x on $\partial(K)$ is moved by f. Thus, $d(f(x), x) > 0$ on $\partial(K)$. Moreover, since K is convex, for all x, the line joining x to $f(x)$ points inward on K. Now, since K is compact, the Stone Weierstrass theorem enables us to uniformly approximate f by a polynomial p in two variables (see [10]). Let $\epsilon > 0$. Then for all $x \in K$, $\|f(x) - p(x)\| < \epsilon$. Consider the C^1 vector field $p(x) - x = g(x)$ on $\partial(K)$. Since K is compact taking ϵ sufficiently small, it follows that $d(p(x), x) > 0$ everywhere on $\partial(K)$. Since K is convex, this vector field always points inward for each $x \in$

---

[2]Actually, it has a deficiency in comparison to the actual Brouwer fixed point theorem in addition to only working when $n = 2$, namely, since the Brouwer theorem is a topological result, it works for any K homeomorphic to the closed n disk and need not be convex This is because convex or not, the region is *homeomorphic* to the disk!

$\partial(K)$ so that it is positively invariant. By the Poincare–Bendixson theorem, K must contain an equilibrium point or a closed orbit. By Corollary 8.1.9, even in the closed orbit case, K must contain an equilibrium point x_0, so $p(x_0) = x_0$ and so $d(f(x_0), x_0) < \epsilon$. Thus, for any small $\epsilon > 0$, there is a point $x_\epsilon \in K$ with $d(f(x_\epsilon), x_\epsilon) < \epsilon$. A simple compactness argument, which we leave as an exercise to the reader, then tells us that there is some point $x \in K$ such that $f(x) = x$. $\qquad\square$

Chapter 9

Matrix Differential Equations and the Matrix Exponential Function

9.1 The Matrix Exponential Function

In order to study matrix differential equations, it will be necessary to define the matrix exponential function Exp on the $n \times n$ complex matrices, i.e., on the space $M_n(\mathbb{C})$ (and then by restriction to $M_n(\mathbb{R})$) rather than on $M_n(\mathbb{R})$ as it takes no more effort and will actually be helpful. These spaces are linear associative algebras over $\mathbb{C}$ and $\mathbb{R}$, respectively, with matrix addition, scalar multiplication and multiplication of matrices as the operations. Due to its importance this function, it will prove to be quite useful in many ways, particularly in applications of linear algebra to ordinary differential equations (ODEs). However, defining Exp will require not only algebra but also some analysis. First, let us consider linear operators T on $\mathrm{End}_k(V)$, where $k = \mathbb{R}$ or $\mathbb{C}$ and $V = \mathbb{C}^n$ or $\mathbb{R}^n$, respectively. Thus, $T(v + w) = T(v) + T(w)$ and $T(cv) = cT(v)$ for all $v, w \in V$ and $c \in k$. On the vector space V, we place any one of its various equivalent norms. We ask the reader to verify that such an operator is continuous on V if and only if there is some positive constant c so that for all $x \in V$, $\|T(x)\| \leq c \cdot \|x\|$. Taking the infimum of such c to be the operator norm, $\|T\|$, of T, we have

$$\|T(x)\| \leq \|T\| \cdot \|x\|$$

Such linear operators are called *bounded* and $\|T\|$ is called the *operator norm*. With this norm and the associated distance

$d(S,T) = \|S - T\|$, $M_n(\mathbb{C})$ and $M_n(\mathbb{R})$ are complete metric spaces. The completeness meaning that the Cauchy criterion is satisfied. The Cauchy criterion states that if a sequence of operators T_n is Cauchy (meaning that given $\epsilon > 0$ there is an integer p so that if $n, m \geq p$, then $d(T_n, T_m) < \epsilon$), then this sequence has a convergent subsequence.[1]

Corollary 9.1.1. *Let S and T be bounded linear operators on V. Then ST is also a bounded linear operator and*

$$\|ST\| \leq \|S\| \cdot \|T\|$$

Proof.   Of course, ST is linear. Moreover,

$$\|ST(x)\| \leq \|S\| \cdot \|T(x)\| \leq \|S\| \cdot \|T\| \cdot \|x\|. \qquad \square$$

This norm inequality makes these linear associative algebras into what are called *Banach algebras*. We now use this simple but important inequality to study the *exponential map*.

As we shall see, the series defining Exp is absolutely convergent and uniformly convergent on compact subsets of $M_n(\mathbb{C})$, and therefore for a complex matrix A, we will define a $\mathbb{C}$ linear operator called Exp(A) by this convergent power series. Exp will map from $M_n(\mathbb{C})$ to itself and evidently since the coefficients are real, Exp$(M_n(\mathbb{R}))$.

Thus, for a fixed, $A \in M_n(\mathbb{C})$ we define Exp(A) by the power series for exp applied to matrices A, namely,

$$\mathrm{Exp}\, A = \sum_{k=0}^{\infty} \frac{A^k}{k!}$$

In particular, Exp$(0) = I$. Now, we will show this series is absolutely convergent and uniformly convergent on compact subsets, therefore it defines an entire holomorphic function

$$\mathrm{Exp}\colon M_n(\mathbb{C}) \to M_n(\mathbb{C})$$

---

[1]Augustin Cauchy (1789–1857) made enumerable contributions to real and complex analysis.

(see [9]). To see this, consider the finite partial sums $\sum_{k=0}^{m} \frac{A^k}{k!}$. Using the Banach algebra inequality to estimate the norm, we get

$$\left\| \sum_{k=n}^{m} \frac{A^k}{k!} \right\| \leq \sum_{k=n}^{m} \frac{\|A\|^k}{k!}.$$

Since the series $\sum_{k=0}^{\infty} \frac{\|A\|^k}{k!}$ converges, it is the series for $e^{\|A\|}$ and it follows that what it dominates must also converge. Thus, $\operatorname{Exp} A$ converges absolutely and uniformly on compact subsets of $M_n(\mathbb{C})$, and moreover, $\|\operatorname{Exp} A\| \leq e^{\|A\|}$.

The following fact is fundamental and explains the importance of the matrix exponential function. Of course, we get it for free when we deal with exp.

Theorem 9.1.2. *If A and B commute, then* $\operatorname{Exp}(A + B) = \operatorname{Exp}(A) \operatorname{Exp}(B)$.

Proof. Now, $\operatorname{Exp}(A) \operatorname{Exp}(B) = \sum_{k=0}^{\infty} \frac{A^k}{k!} \sum_{l=0}^{\infty} \frac{A^l}{l!} = \sum_{l,k}^{\infty} \frac{1}{k!l!} A^k B^l$. On the other hand, $\operatorname{Exp}(A + B) = \sum_{p=0}^{\infty} \frac{(A+B)^p}{p!}$. Since A and B commute, the binomial theorem tells us that

$$\frac{(A + B)^p}{p!} = \sum_{j=0}^{p} \frac{A^j}{j!} \frac{B^{p-j}}{(p - j)!}.$$

Due to the absolute convergence of the series, the Weierstrass rearrangement theorem tells us that these are equal (see Appendix D). $\square$

We remark, and the reader should check that the binomial theorem holds in any commutative ring with identity.

From this, we see that $\operatorname{Exp}(A)$ is always invertible. In fact, since A and $-A$ certainly commute and $\operatorname{Exp}(0) = I$, we get

Corollary 9.1.3. $\operatorname{Exp}(A)^{-1} = \operatorname{Exp}(-A)$.

Thus, $\operatorname{Exp} : M_n(\mathbb{C}) \to \operatorname{GL}_n(\mathbb{C})$ and $\operatorname{Exp} : M_n(\mathbb{R}) \to \operatorname{GL}_n(\mathbb{R})$. Here, $\operatorname{GL}_n(k)$ is the group of invertible $n \times n$ square matrices with coefficients from the field k, where $k = \mathbb{C}$ or $\mathbb{R}$.

Moreover, since tA and sA commute for all real t and s, it follows that $\operatorname{Exp} t(A + B) = \operatorname{Exp}(tA) \operatorname{Exp}(sB)$. That is, for each A,

$f(t) = \text{Exp}(tA)$ is a multiplicative group homomorphism. Our next result shows that Exp itself commutes with conjugation.

Proposition 9.1.4. *Let $A \in M_n(\mathbb{C})$ and $P \in \text{GL}_n(\mathbb{C})$, then*

$$P \text{Exp}(A)P^{-1} = \text{Exp}(PAP^{-1}).$$

Proof. This follows immediately from the fact that conjugation by $P \in \text{GL}_n(\mathbb{C})$ is an algebra automorphism. Hence, for any positive integer j and any constant $c \in \mathbb{C}$, we have $PcA^jP^{-1} = c(PAP^{-1})^j$. Therefore, for any polynomial $f(A)$, we have $Pf(A)P^{-1} = f(PAP^{-1})$. Now, suppose $f(A)$ is an everywhere convergent power series. Taking limits as the partial sums tend to $f(A)$ tells us that $Pf(A)P^{-1} = f(PAP^{-1})$. $\qquad\square$

That conjugation commutes with Exp allows us to recapture Liouville's[2] formula, namely, for any complex $n \times n$ matrix A,

$$\det(\text{Exp } A) = e^{\text{tr}(A)}.$$

In particular, $\det(\text{Exp}(A)) = 1$ if and only if $\text{tr}(A) = 0$ and by replacing A by tA, we see that this formula holds for all real t,

$$\det(\text{Exp}(tA)) = e^{\text{tr}(tA)} = e^{t\,\text{tr}(A)}.$$

Proof. As we noted earlier, this is because when A is triangular with eigenvalues $\lambda_1, \ldots, \lambda_n$, then a direct calculation shows that $\text{Exp } A$ is also triangular with eigenvalues $e^{\lambda_1}, \ldots, e^{\lambda_n}$. Hence,

$$\det(\text{Exp } A) = e^{\lambda_1} \cdots e^{\lambda_n} = e^{\lambda_1 + \cdots + \lambda_n} = e^{\text{tr}(A)}.$$

Since $\mathbb{C}$ is algebraically closed, we can now apply the third Jordan canonical form to get a P so that PAP^{-1} is triangular (see [4]). Then $P \text{Exp}(A)P^{-1} = \text{Exp}(PAP^{-1})$. The determinant, respectively the

---

[2] Joseph Liouville (1809–1882) was a French mathematician who made important contributions to number theory, complex analysis, differential geometry, topology and differential equations. He founded the *Journal de Mathématiques Pures et Appliquées* and was the first to read and to recognize the importance of the unpublished work of Évariste Galois, which was published in his journal in 1846 some 15 years after Galois' death.

trace of a triangular matrix, is the same as that of the non-triangular one, hence the result. $\qquad\square$

We can now get more detail concerning the invertibility of $\mathrm{Exp}(A)$ by understanding the relationship between the eigenvalues of A and those of $\mathrm{Exp}(A)$.

Proposition 9.1.5. *If λ is an eigenvalue of A, then e^λ is an eigenvalue of $\mathrm{Exp}(A)$.*

Proof. By the Jordan canonical form (see [4] Chapter 6), we can put A in triangular form over $\mathbb{C}$. That is, there is an invertible operator P so that PAP^{-1} is triangular, say with eigenvalues $\lambda_1, \ldots, \lambda_n$ and as we observed, $\mathrm{Exp}(PAP^{-1})$ is also triangular with eigenvalues $e^{\lambda_1}, \ldots, e^{\lambda_n}$. $\qquad\square$

Here is a connection of some of this to differential equations. More will follow.

Consider the differential equation $\phi'(t) = A\phi(t)$, where ϕ takes values in the complex $n \times n$ matrices and A is a fixed matrix. Here, t ranges over all of $\mathbb{R}$, ϕ is smooth and our initial condition is $\phi(0) = I$. This is a first-order linear matrix differential equation with constant coefficients (or a system of n^2 such numerical equations) and hence has a unique *global* solution (see Appendix B). Since $t \mapsto \mathrm{Exp}(tA)$ has these same properties, $\phi(t) = \mathrm{Exp}(tA)$ for all real t.

In particular, we get a converse of the fact that $\phi(t) = \mathrm{Exp}(tA)$ is a group homomorphism, for suppose $\phi(t)$ is any homomorphism. Since for all s and $t \in \mathbb{R}$, $\phi(s + t) = \phi(s)\phi(t)$. Differentiating with respect to s at $s = 0$ gives $\phi'(t) = \phi'(0)\phi(t)$ for all real t. Since $\phi(0) = I$, it follows from the previous paragraph that $\phi(t) = \mathrm{Exp}(tA)$ for some matrix A.

We remark that $\frac{d}{dt}(\mathrm{Exp}\,tA)|_{t=0} = A$ means A is the tangent vector to ϕ at $t = 0$. From this, it follows easily that the derivative of Exp at 0 is the identity, $D_{\mathrm{Exp}}(0) = I$, and therefore, the Jacobean, $\det(D_{\mathrm{Exp}}(0)) \neq 0$. Since Exp is smooth, the inverse function theorem tells us that Exp itself is a local diffeomorphism of a neighborhood of 0 in $M_n(\mathbb{C})$ with a neighborhood of I in $\mathrm{GL}_n(\mathbb{C})$ (which is connected), respectively a neighborhood of 0 in $M_n(\mathbb{R})$ with a neighborhood of I in $\mathrm{GL}_n(\mathbb{R})_+$ (the connected component of I). This is a typical argument using *differential topology* and it only yields *some* neighborhood. We can do better by power series.

From ordinary calculus of 1 real variable, we know that when $|x - 1| < 1$, $\log(x)$ is given by the convergent power series

$$\sum_{n \geq 1} \frac{(-1)^{n-1}}{n} (x - 1)^n,$$

and that if $|x - 1| < 1$, $\exp(\log(x)) = x$. It follows that in the abstract ring of power series, $\mathbb{R}[x]$, the same must hold, i.e., for *all* x, the composition of these power series yields x. Now, the convergence of the log real power series for $|x - 1| < 1$ enables us to define for a matrix $A \in M_n(\mathbb{C})$ with $\|A - I\| < 1$, the Log function by convergent matrix power series, namely,

$$\mathrm{Log}(A) = \sum_{n \geq 1} \frac{(-1)^{n-1}}{n} (A - I)^n.$$

As a convergent power series, this function is smooth and since the

$$\exp(\log(x)) = x$$

in the abstract power series ring holds, the same must be true in the matrix power series wherever both Exp and Log converge. Since Exp converges everywhere and Log converges when $\|A - I\| < 1$, it follows that $\mathrm{Exp}(\mathrm{Log}(A)) = A$ whenever $\|A - I\| < 1$. Thus, Exp is 1:1 on $\mathrm{Log}(\mathcal{N}(0,1))$, where $\mathcal{N}(p, \delta)$ is the open ball in $M_n(\mathbb{C})$, or some subset, centered at p of radius δ. Of course, by exactly the same argument, $\mathrm{Log}(\mathrm{Exp}(B)) = B$ whenever $\|\mathrm{Exp}(B) - I\| < 1$.

Since Exp and Log are smooth functions, by continuity of Exp, there is a $\delta > 0$ so that $\mathcal{N}(0, \delta) \subseteq \mathrm{Exp}^{-1} \mathcal{N}_{\mathrm{GL}(n,\mathbb{C})}(I, 1)$. Applying Exp tells us that

$$\mathrm{Exp}(\mathcal{N}(0, \delta)) \subseteq \mathcal{N}_{\mathrm{GL}(n,\mathbb{C})}(I, e^\delta - 1),$$

and taking $\delta = \log(2)$, we see

$$\mathrm{Exp}(\mathcal{N}(0, \log(2))) \subseteq \mathcal{N}_{\mathrm{GL}(n,\mathbb{C})}(I, 1),$$

thus giving an explicit and pretty large neighborhood of zero where Exp is invertible.

We conclude this section with an exercise.

Exercise 9.1.6. Prove that for an $n \times n$ matrix A and for m a positive integer, $\mathrm{Exp}\, A = \lim_{m \to \infty} (I + \frac{A}{m})^m$.

9.1.1 Extending Liouville's theorem to the case of variable coefficients

Here, we extend Liouville's theorem above for matrix ODEs to the case of variable coefficients. That is, we consider the matrix differential equation,

$$\frac{dX(t)}{dt} = A(t) \cdot X(t), \qquad (9.1)$$

where $X(t)$ is the unknown matrix-valued function of t and $A(t) = (a_{ij}(t))$ are the matrix of coefficients. Here, t lies in some connected interval I containing 0 in $\mathbb{R}$ which could be a finite, or half infinite interval, or $\mathbb{R}$ itself. As initial condition, we continue to take $X(0) = I$. We consider the linear homogeneous ODE (9.1), where $X(t)$ is the unknown path in the space of $n \times n$ real or complex matrices and $A(t)$ is a given smooth $n \times n$-matrix valued function of $t \in I$. In coordinates, this is a system of n^2 ODEs in n^2 unknowns whose coefficients are $a_{ij}(t)$ and so always has local solutions.

Liouville's theorem states and we will prove that if $A(t)$ is real analytic and $X(t)$ satisfies the differential equation above, then $\det(X(t))$ satisfies its own (numerical) differential equation.

Theorem 9.1.7. *Let $A(t)$ be a real analytic function of t and suppose $X(t)$ satisfies the matrix ODE with initial condition as above. Then also on our interval I, $\frac{d \det(X(t))}{dt} = \mathrm{tr}(A(t)) \det(X(t))$, with initial condition $\det(X(0)) = 1$.*

We shall prove this using two lemmas.

Lemma 9.1.8. *Let A be a fixed $n \times n$ real matrix. Then*

$$\det(I + tA) = 1 + t\,\mathrm{tr}(A) + O(t^2), t \in \mathbb{R}.$$

In particular, $\frac{d}{dt}|_{t=0}(\det(I + tA + O(t^2))) = \mathrm{tr}(A)$.

Proof.   Let $\lambda_1, \ldots, \lambda_n$ be the eigenvalues of A counted with multiplicity. By the third Jordan form, A can be put in triangular form over $\mathbb{C}$ with λ_i on the diagonal. Therefore, $I + tA$ is also in triangular form with $1 + t\lambda_i$ on the diagonal. Hence, $\det(I + tA) = \prod_{i=1}^{n}(1 + t\lambda_i)$. This last term is $1 + t\operatorname{tr}(A) + O(t^2)$.   □

Lemma 9.1.9. *If the coefficient function $A(t)$ is analytic, then the places where $\det(\frac{dX(t)}{dt}) \neq 0$ on an open dense set in I.*

Proof.   Using the local theory, one knows that since $A(t)$ is analytic as a solution, $X(t)$ is locally analytic and therefore (real) analytic. Hence, so is $\frac{dX(t)}{dt}$ and since $\det$ is a polynomial, the same is true of $\det(\frac{dX(t)}{dt})$. The conclusion then follows, for example, from Corollary (6.4.4) of [9].   □

Proof of the theorem is as follows.

Proof.   Choose $t_0 \in I$ to be in the open dense set I_0 of Lemma 9.1.9 as well. Then the linear approximation for $X(t)$ at t_0 is

$$X(t) = X(t_0) + (t - t_0)X'(t_0) + O((t - t_0)^2).$$

Since $X(t)$ is a solution to the ODE, the right-hand side is $X(t_0) + (t - t_0)A(t_0)X(t_0) + O((t - t_0)^2)$. Due to our choice of t_0, $X'(t_0)$ is invertible. Hence, this last term is $(I + (t - t_0)A(t_0) + O(t - t_0)X(t_0)^{-1})X(t_0)$ which is just $(I + (t - t_0)A(t_0) + O(t - t_0))X(t_0)$ and so $X(t) = (I + (t - t_0)A(t_0) + O(t - t_0))X(t_0)$. Taking determinants of both sides yields

$$\det(X(t)) = \det(I + (t - t_0)A(t_0) + O(t - t_0)^2)\det(X(t_0)).$$

By Lemma 9.1.8, the derivative at t_0 of $\det(I + (t - t_0)A(t_0) + O(t - t_0)^2)$ is $A(t_0)$. Thus,

$$\frac{d}{dt}(\det(X(t))) = \operatorname{tr}(A(t))\det(X(t))$$

for all $t \in I_0$. By continuity, this then holds for all $t \in I$.   □

We now consider the Liouville equation itself. This recaptures what we did above in the case when the coefficients are constants,

$A(t) \equiv A$. For then the solution is global; it is $X(t) = X(0)\operatorname{Exp} tA$, $t \in I$. In this case, $I = \mathbb{R}$, and our ODE is $\frac{dX(t)}{dt} = A \cdot X(t)$ with initial condition, $X(0) = I$. Then as we noted, $X(t) = \operatorname{Exp} tA$ and so for all t, $\det(\operatorname{Exp} tA) \equiv \exp(t\operatorname{tr}(A))$. From this, we reencounter the *Wronskian*.

Corollary 9.1.10. $\frac{\det(\operatorname{Exp}(tA))}{dt} = \operatorname{tr}(A)e^{t\operatorname{tr}(A)}$.

Thus, $W'(t) = \frac{\det(\operatorname{Exp}(tA))}{dt} = \operatorname{tr}(A)e^{t\operatorname{tr}(A)} = \operatorname{tr}(A)W(t)$. So, as before, the Wronskian satisfies its own numerical first-order ODE whose solution we already know from Chapter 1. Here we write $M_n(\mathbb{C}) = M(n,\mathbb{C})$ etc.

9.1.2   The range of Exp

For an $A \in M(n,\mathbb{R})$, or $M(n,\mathbb{C})$, we ask what can be said about the range of $\operatorname{Exp}(tA)$ as t varies over $\mathbb{R}$? The path starts out at I when $t = 0$ and $\operatorname{Exp}(tA)$ is always invertible, so as we know, it stays in $\operatorname{GL}(n,\mathbb{R})$ or $\operatorname{GL}(n,\mathbb{C})$. In the real case, one can say a little more. Since $\det(\operatorname{Exp} tA) = e^{t\operatorname{tr}(A)}$, which in the real case is positive for all real t, we see that $\operatorname{Exp}(tA) \in \operatorname{GL}(n,\mathbb{R})_+$, the real matrices of positive determinant for all real t. Thus, $\operatorname{Exp} : M(n,\mathbb{R}) \to \operatorname{GL}(n,\mathbb{R})_+$ and $\operatorname{Exp} : M(n,\mathbb{C}) \to \operatorname{GL}(n,\mathbb{C})$. We ask if either of these maps is surjective.

In the complex case, Exp is indeed surjective because when $A \in M(n,\mathbb{C})$, it is triangularizable by the Jordan canonical form (see [4] in Chapter 6), so since Exp commutes with conjugation, we may as well assume A is itself in triangular form. Indeed, grouping like eigenvalues together, we can write A in block diagonal form as a sum of blocks, $A_j = N_j + \lambda_j I$, where N_j is nil triangular and $\lambda_j I$ is scalar, the order of each block being the multiplicity of the respective eigenvalue. Restricting Exp to each of these blocks, since N_j commuted with $\lambda_j I$, $\operatorname{Exp}(A_j) = \operatorname{Exp}(N_j)(e^{\lambda_j}I)$. As is well known (both in the real or complex case), $\operatorname{Exp}(N_j) = U_j$, where U_j are the unitriangular matrices of that order. Since $\exp : \mathbb{C} \to \mathbb{C}^\times$ is surjective, Exp is surjective on each of the blocks and therefore maps surjectively $M(n,\mathbb{C}) \to \operatorname{GL}(n,\mathbb{C})$.

Of course, this argument fails over $\mathbb{R}$. Indeed, already when $n = 2$, the statement is false; $\text{Exp} : M(n, \mathbb{R}) \to \text{GL}(n, \mathbb{R})_+$ is not surjective. Let B be a 2×2 triangular, *but not diagonalizable* matrix with both diagonal elements equal to -1 and suppose $\text{Exp}(A) = B$. Let λ and μ be the eigenvalues of A. Then $e^\lambda = -1 = e^\mu$. Clearly, neither λ nor μ are real since the range of exp is positive over $\mathbb{R}$. Therefore, $\lambda = a + bi$ and $\mu = a - bi$, where $b \neq 0$. So, $e^a(\cos(b) + i\sin(b)) = -1$. Therefore, since $b \neq 0$, $b = \pi$ and $a = 0$. This means that the eigenvalues of A are $\pm \pi i$. Hence, A is diagonalizable. Therefore, $\text{Exp}(A) = B$ is also diagonalizable, a contradiction.

This counter example suggests the following.

Proposition 9.1.11. *In the real case, if $B \in \text{GL}(n, \mathbb{R})_+$, there is always an $A \in M_n(\mathbb{R})$ with $\text{Exp}(A) = B^2$.*

Proof.   This is because if

$$B = \begin{bmatrix} -1 & 0 \\ * & -1, \end{bmatrix},$$

where $* \neq 0$. Then

$$B^2 = \begin{bmatrix} 1 & 0 \\ ** & 1 \end{bmatrix}.$$

This unipotent matrix is the exponential image of a real nilpotent matrix. see [1].

To complete the proof of the proposition, we use the Jordan form in the case of the real matrix A which is block triangular with 2×2 and 1×1 blocks. These last are the real eigenvalues of A. It would be sufficient to prove the proposition in the case of a 2×2 block and in the case of all real eigenvalues. The former having already been shown to be exponential, i.e., $\text{Exp}(A) = B$ even if B were not a square, since then $\text{Exp}(2A) = B^2$. Thus, we are reduced to the case where all eigenvalues of A are real. Since we know $\det(A) > 0$, there are no zero eigenvalues (and of course, any negative eigenvalues must occur an even number of times, but this doesn't matter). Then $\text{Exp}(2A) = B^2$, where B^2 is triangular with all positive eigenvalues (the squares of those of B). The proof can then proceed by induction on n.   □

Exercise 9.1.12. Prove the following:

1. If A is a real skew symmetric matrix, then $\mathrm{Exp}(A)$ is orthogonal.
2. If A is a skew Hermitian matrix, then $\mathrm{Exp}(A)$ is unitary.
3. If A is a triangular matrix, then $\mathrm{Exp}(A)$ is triangular.
4. If A is a nilpotent matrix, then $\mathrm{Exp}(A)$ is unipotent.

Exercise 9.1.13. Prove that the $ax + b$ group is exponential, i.e., the exponential map, Exp from

$$x = \begin{pmatrix} a & b \\ 0 & 0 \end{pmatrix},$$

where $a, b \in \mathbb{R}$ maps to

$$g = \begin{pmatrix} \alpha & \beta \\ 0 & 1 \end{pmatrix},$$

where $\alpha > 0$ and $\beta \in \mathbb{R}$ are arbitrary. That is, Exp is surjective.

Observe that what we have done earlier with the exponential map in dimension $n \geq 2$ proves that the analogous statement in higher dimensions fails. That is, the map from

$$X = \begin{pmatrix} A & \mathbf{b} \\ 0 & 0 \end{pmatrix},$$

where $A \in M_n(\mathbb{R})$, $n \geq 2$ and $\mathbf{b} \in \mathbb{R}^{\mathbf{n}}$ maps to

$$g = \begin{pmatrix} \alpha & \beta \\ 0 & 1 \end{pmatrix},$$

where $\alpha \in M_n(\mathbb{R})$ with $\det(\alpha) > 0$ and $\beta \in \mathbb{R}^{\mathbf{n}}$, is not surjective.

This shows that, in general, given a *connected* Lie group G and its Lie algebra, $\mathfrak{g}$ the exponential map $\exp : \mathfrak{g} \to G$ is not surjective. However, in any connected Lie group, G, any element $g \in G$ is a product of two exponentials, namely, $g = \exp_G(x) \exp_G(Y)$ for some X, Y in its Lie algebra $\mathfrak{g}$ (see [11]).

Exercise 9.1.14. Prove the Leibnitz formula for determinants,

$$\frac{\det X(t)}{dt} = \sum_{i=1}^{n} \det(X_{i,j}^*),$$

where the matrix $(X_{i,j}^*)$ is just X, except that in the ith row, we have the derivatives, $X_{i,j}'$, rather than $X_{i,j}$. *Suggestion*: Using the ODE just above, it follows that each

$$X_{i,k}' = \sum_{j=1}^{n} a_{i,j} Xj, k, \quad \text{for } i, k = 1, \ldots, n$$

so that the entire ith row satisfies

$$(X_{i,1}', \ldots, X_{i,n}') = \sum_{j=1}^{n} a_{i,j}(X_{i,1}, \ldots, X_{i,n}).$$

Remark 9.1.15. As the reader has no doubt observed, in the early chapters of this book, a number of differential equations were solved by an adroit change of variables. This process came to be formalized in the late 19th and early 20th century as follows: Suppose $\frac{dX}{dt} = F(t, X)$ with $X(0) = X_0$ is a system of differential equations in an open subset $U \subseteq \mathbb{R}^n$, where F is a smooth function and $X(t)$ the sought after solution curve in U. We can try to change variables by composing with a bijective diffeomorphism $\phi : U \to U$ (for example, by going from rectangular coordinates to polar coordinates in $U = \mathbb{R}^2 \setminus \{(0,0)\}$). The set of all such change of variables is a group, $\mathcal{G}$. Moreover, suppose the differential equation was invariant under a continuous group of transformations, G, called the symmetry group of the equation. Such a group is defined by its elements preserving some quantity, i.e., by a subgroup $G \subseteq GL(n, \mathbb{R})$ within $\mathcal{G}$ which preserve that quantity. (For example, $G = SL(n, \mathbb{R})$ consists of the matrices of det 1, while $G = O(n, \mathbb{R})$ consists of all the orthogonal transformations.) Such a G is a Lie group. Given G, one considers 1 parameter subgroups of it parametrized by $\mathbb{R}$ so that $\phi(0, *)$ denotes the identity transformation, i.e., for $X \in U$, $\phi(0, X) = X$ and for all $s, t \in \mathbb{R}$ and $X \in U$, $\phi(s, \phi(t, X)) = \phi(s + t, X)$. Then there will exist an essentially unique $A \in \mathfrak{g}$, the Lie algebra of G, so that $\mathrm{Exp}(tA) = \phi(t, *)$. (Actually, the line through A is what is unique.)

The study of these ODEs will then devolve to the properties of G and its Lie algebra. Such a study for both ODEs and PDEs is carried out, for example, in [6].

9.2 The Matrix Exponential Function and Linear ODEs with Constant Coefficients

Since

$$\text{Exp}(tA) = \sum_{n=0}^{\infty} t^n A^n,$$

in particular, if we want to calculate $\frac{d}{dt}(\text{Exp}(tA))$, we can differentiate this series term by term (see Appendix D). As a corollary, we can solve the linear homogeneous ODE with constant coefficients,

$$\frac{dX(t)}{dt} = AX(t), \tag{9.2}$$

$$X(0) = I, \tag{9.3}$$

where $X(t)$ is the unknown path in the space $M_n(\mathbb{R})$ which is defined for all $t \in \mathbb{R}$. Here, A is a fixed $n \times n$ real matrix (when $n = 1$, this is the ODE which defines the exponential function). In coordinates, we see that this is a system of n^2 numerical ODEs in n^2 unknowns, whose coefficients are $a_{i,j}$ and so it always has local solutions, but here since the coefficients are constants, the solution is global (see Appendix B).

However, since we are interested in explicit solutions, we don't reason in this way. Term by term differentiation of the power series (see Appendix D), $\text{Exp}\, tA$ tells us that

$$\frac{d}{dt}(\text{Exp}(tA)) = (\text{Exp}(tA)) \cdot A.$$

Thus, $X(t) = \text{Exp}(tA)$ satisfies the differential equation (9.2) for all $t \in \mathbb{R}$, and of course, $X(0) = I$. Conversely, if $X(t)$ satisfies the ODE, then using the initial condition and term by term integration (see Appendix D), it follows that $X(t) = \text{Exp}(tA)$.

There is another related ODE to the one just above. This is the linear homogeneous ODE with constant coefficients,

$$\frac{d\mathbf{x}(\mathbf{t})}{dt} = A\mathbf{x}(\mathbf{t}), \tag{9.4}$$

where $\mathbf{x}(\mathbf{t})$ is an unknown path in $\mathbb{R}^n$ defined for all $t \in \mathbb{R}$ and A is a fixed $n \times n$ real matrix. Let $\mathbf{x}(\mathbf{t}) = \mathrm{Exp}(\mathbf{t}A)(\mathbf{x_0})$, where $\mathbf{x_0}$ is a fixed vector in $\mathbb{R}^n$. Then just as before, a direct calculation yields

$$\frac{d\mathbf{x}(\mathbf{t})}{dt} = A\,\mathrm{Exp}(tA)(\mathbf{x_0}) = \mathbf{A}\mathbf{x}(\mathbf{t}).$$

Hence, $\mathbf{x}(\mathbf{t}) = \mathrm{Exp}(\mathbf{t}A)(\mathbf{x_0})$ is a solution, where $\mathbf{x_0}$ is any fixed vector in $\mathbb{R}^n$. Since $\mathrm{Exp}(tA)$ is a linear operator, the space of solutions is an $\mathbb{R}$-linear vector space of dimension n, indexed by the various $\mathbf{x_0} \in \mathbb{R}^n$. This map is an $\mathbb{R}$-linear isomorphism. Hence, if $\{\mathbf{x_1}, \dots, \mathbf{x_n}\}$ is a basis of the $\mathbb{R}$ vector space $\mathbb{R}^n$, then any solution to (9.4) is of the form $\mathbf{x}(\mathbf{t}) = \sum_{\mathbf{i=1}}^{\mathbf{n}} \mathbf{c_i}\,\mathrm{Exp}(\mathbf{t}A)(\mathbf{x_i})$ for unique constants $c_i \in \mathbb{R}$.

In particular, if $\mathbf{x}(\mathbf{t})$ satisfies the ODE and vanishes at a point $(\mathbf{x}(\mathbf{t_0}) = \mathbf{0})$, then $\mathbf{x}(\mathbf{t}) = \mathbf{0}$ for all t. This is because A and $\mathrm{Exp}(tA)$ commute, $\mathrm{Exp}(tA)$ is invertible and the set of A which are invertible is dense in $M_n(\mathbb{R})$.

Of course, getting n linearly independent solutions for the vector ODE is the same as getting a solution to the matrix ODE since each is equivalent to the associated system of n^2 numerical ODEs.

Finally, we remark that (9.4) and what follows could be used to give alternative proofs of the results on constant coefficient systems of 2×2 linear systems of ODEs of Chapter 7. This concludes our work on (homogeneous) matrix ODEs with constant coefficients.

9.3   Homogeneous versus Non-Homogeneous Equations and the Variation of Constants

Given a homogeneous system with constant coefficients, i.e., when A is a *constant* $n \times n$ real matrix, we call the one parameter group determined by A, i.e., $\mathrm{Exp}\,tA$, where $t \in \mathbb{R}$ its *fundamental matrix*.

Let I be a fixed closed interval on $\mathbb{R}$. Here, we consider possibly variable coefficients

$$\frac{d\mathbf{x}(t)}{dt} = A(t)\mathbf{x}(t),$$

where $\mathbf{x}(t)$ is an unknown path in $\mathbb{R}^n$, $A(t)$ is an $n \times n$ smooth real matrix function of t defined on I, and we compare solutions with those of the non-homogeneous equation

$$\frac{d\mathbf{x}(t)}{dt} = A(t)\mathbf{x}(t) + \mathbf{b}(t), \tag{9.5}$$

where $b(t)$ is a smooth vector-valued function defined for all $t \in I$.

As we have already observed, there exists a unique solution to (9.5) satisfying the initial condition $\mathbf{x}(t_0) = \mathbf{v}$ (see Appendix B). The same is true of (9.5). Here, the uniqueness can be seen directly since if we had two different solutions on I, their difference would be a solution to (9.5) on I and it would satisfy $\mathbf{x}(t_0) = \mathbf{0}$ and hence by equation (9.5), the difference would be constant and because of this, initial condition is therefore zero.

Knowing the fundamental matrix $\mathrm{Exp}(tA) = \Phi(t)$, then for $t \in I$, we can consider

$$\phi(t) = \Phi(t) \int_\tau^t \Phi^{-1}(s)b(s)ds. \tag{9.6}$$

We shall now show ϕ is the solution to (9.6) satisfying $\phi(\tau) = 0$.

Proof. First, since $\int_\tau^\tau \Phi^{-1}(s)b(s)ds = 0$, we see that $\phi(\tau) = 0$. Moreover, we also know equation (9.6) also has a solution on I. Indeed, it has a unique solution on I satisfying $\phi(\tau) = 0$. Now, let c be a fixed vector, or parameter on $\mathbb{R}^n$. What would it take for $\phi = \Phi(c)$ to be a solution to (9.6)? If ϕ were such a solution, then $\phi' = \Phi'(c) + \Phi(c') = A\Phi(c) + \Phi(c')$. But since as we know $\Phi(c)$ is a solution to (9.6), this is $A\phi + \Phi(c')$ and since ϕ is supposed to be a solution to (9.6), this is $A\phi + b$ for some b. Therefore, $\Phi(c')$ must be b. That is, $c' = \Phi^{-1}b$. Thus, (9.6) is always solvable by taking, for $t \in I$,

$$c(t) = \int_\tau^t \Phi^{-1}(s)b(s)ds.$$

Doing so would force $c(\tau) = 0$. Thus, ϕ is *the* solution of (9.6), satisfying $\phi(\tau) = 0$!

From this, one sees easily that any solution to (9.6) can be obtained by adding an arbitrarily chosen solution of equation (9.6) to this fixed solution to (9.6). Thus, if $\phi(\tau) = \eta$, then

$$\phi(t) = \phi_h(t) + \Phi(t) \int_\tau^t \Phi^{-1}(s)b(s)ds, \qquad (9.7)$$

where ϕ_h is a solution to (9.7).   □

Equation (9.7) is called the *variation of constants formula*.

9.4   The Matrix Exponential Function and Linear ODEs with Variable but Periodic Coefficients: The Floquet–Liapunov Theorem

We now turn to ODEs of the same form as above, but this time we allow variable coefficients which are however *periodic*. That is, we shall assume that there exists a least positive number τ such that for all real t, $A(t + \tau) = A(t)$. The ideas concerning matrix ODEs with periodic coefficients are due independently to both Liapunov and Floquet. Here, we follow Pontrjagin (see [12]) who for some reason neglects to mention Floquet.[3]

Thus, we are interested in differential equations of the form,

$$\frac{dX(t)}{dt} = A(t)X(t),$$

where $A(t)$ is periodic.

Now, as above in the case of constant coefficients, in the first instance, $X(t)$ is an unknown path in $M_n(\mathbb{R})$ and the operation is matrix multiplication, while in the second, the unknown path is in $\mathbb{R}^n$ and the operation is evaluation. We now come to the concepts of a *fundamental matrix* of a solution to (9.4) and *equivalence of solutions* to such an equation.

Definition 9.4.1. Suppose $A(t + \tau) = A(t)$ as above and $\Phi(t)$ is a solution to equation 9.4 over $\mathbb{R}$. If there is a constant invertible

---

[3]Gaston Floquet (1847–1920) was a French mathematician, best known for his work in ODE.

matrix C such that $\Phi(t+\tau) = \Phi(t)C$ for all real t, then we shall call C a *fundamental matrix* of the solution Φ. If Ψ is another solution to (9.4) and D is a fundamental matrix for Ψ, then we say Φ and Ψ are equivalent if there exists a non-singular matrix P such that for all t, $\Psi(t) = \Phi(t)P$.

Proposition 9.4.2.

1. *Every solution to* (9.4) *has a fundamental matrix.* (*It should be understood that the matrices* C *and* D *are constant, when we hold* t *fixed, and therefore are actually functions of* t.)
2. *Furthermore, if* Φ *and* Ψ *are solutions to* (9.4) *with fundamental matrices* C *and* D, *respectively, then* Φ *and* Ψ *are equivalent if and only if* $D = P^{-1}CP$ *for some invertible* P *and* $\Phi = \Psi P$. *Evidently, this is an equivalence relation.*

Proof. We first observe that if $\Phi(t)$ is a solution to (9.4), then so is $\Phi(t+\tau)$. This is because $\Phi(t+\tau)' = A(t+\tau)\Phi(t+\tau) = A(t)\Phi(t+\tau)$. Now, recall the linear homogeneous ODE with constant coefficients above:

$$\frac{d\mathbf{x}(\mathbf{t})}{dt} = A\mathbf{x}(\mathbf{t}),$$

where $\mathbf{x}(\mathbf{t})$ is an unknown path in $\mathbb{R}^n$ defined for all $t \in \mathbb{R}$ where A is a fixed $n \times n$ real matrix. The solution is uniquely determined by specifying $\mathbf{x_0}$, a fixed vector in $\mathbb{R}^n$. Taking it to be the elements of a standard basis of $\mathbb{R}^n$, we see that for t fixed, since $A(t) = A(t+\tau)$, and each of these creates an $\mathbb{R}$-linear transformation of the standard basis, there exists an invertible matrix, C, which takes the transformed basis using $A(t)$ to the one using $A(t + \tau)$. This proves the first statement.

Turning to the second, we know that $\Phi'(t) = \Phi(t)P$. Hence,

$$\Phi'(t + \tau) = \Phi(t + \tau)P = \Phi(t)CP = \Phi(t)'P^{-1}CP,$$

proving the second. □

Now, we consider two equations, $X' = A(t)X$ and $Y' = B(t)Y$, with variable, but periodic coefficients, both of the same period, τ. If they are equivalent, then there exists an invertible linear transformation $S(t)$ such that $Y = S(t)X$ where $S(t)$ is a periodic function of t of period τ which transforms the first ODE into the second. Let $X = \Phi(t)$ and $Y = \Psi(t)$ be respective solutions. Then these have the *same* fundamental matrix. For suppose the first of these has C as its fundamental matrix, then $Y = \Psi(t) = S(t)\Phi(t)$ is a solution of the second equation, and

$$\Psi(t + \tau) = S(t + \tau)\Phi(t + \tau) = S(t)\Phi(t + \tau) = S(t)\Phi(t)C$$
$$= \Psi(t)C.$$

Henceforth, we shall assume that the solutions $X = \Phi$ and $Y = \Psi$ have the same C. That is, $\Phi(t + \tau) = \Phi(t)C$ and $\Psi(t + \tau) = \Psi(t)C$. But this forces $\Phi(t + \tau)\Psi(t + \tau)^{-1} = \Phi(t)\Psi^{-1}(t)$. Thus, $S(t) = \Phi(t)\Psi^{-1}(t)$ and so $\Psi(t) = S(t)\Phi(t)$. Since each of these solutions is uniquely determined by its respective initial condition, what we have just proved together with the Proposition 9.4.2, where we observed that Φ and Ψ are equivalent if and only if $D = P^{-1}CP$ for some invertible P, shows that the conjugacy class of C is a complete set of invariants for equivalence classes of solutions to (9.4).

We remark that so far all statements of this section hold for matrix ODEs over either $\mathbb{R}$ or $\mathbb{C}$. However, we now come to a point which will require that we treat these cases separately.

Theorem 9.4.3. *Any matrix ODE of the form* (9.4) *where for all real t, $A(t) = A(t + \tau)$ is equivalent to one with constant coefficients, $Y' = BY$. In general, this B is complex. However, if we start out with all real data (and then of course we would want the "B" to be real and periodic), we will need "B" $= B_1$, a constant real matrix which actually has period 2τ. In this way, when $Y' = B_1Y$, the matrix $S(t)$ will also be real.*

Proof.   As we already know from previous results in this chapter, if C is a fundamental matrix of some solution Φ of (9.4), then there exists a complex matrix B so that $\mathrm{Exp}(\tau B) = C$. We shall prove the ODE $Y' = BY$ is equivalent to a periodic ODE. Now,

$$\mathrm{Exp}[(t + \tau)B] = \mathrm{Exp}(tB)\,\mathrm{Exp}(\tau B) = \mathrm{Exp}(tB)C.$$

Since the fundamental matrices of these equations coincide, this proves the result in the complex case.

Now, let $A(t)$ be periodic in t and real. As before, Φ is a real solution with C as its fundamental matrix. Then $\Phi(t + \tau) = \Phi(t)C$ with C real since Φ is real. But then $\Phi(t + 2\tau) = \Phi(t + \tau)C = \Phi(t)C^2$. As we know by Corollary 9.1.11, there exists a real B_1 such that $\text{Exp}(2\tau B_1) = C^2$. Since the fundamental matrices of the two solutions are both C^2 and so coincide, the proof in the real case proceeds as before. $\square$

Finally, we give a practical sufficient condition for the domain of a solution, Φ, of the periodic ODE, $X' = A(t)X$ of period 2τ to be defined throughout $[0, \infty)$, thus making the qualitative study of Chapter 5 applicable here. As an initial data point, we take $\Phi(0) = I$ and for convenience work over $\mathbb{R}$. Here, we shall prove that there exist positive constants R and d such that

$$\|\Phi(t)\| \leq \text{Re}^{dt} \quad \text{for every } t \in [0, \infty).$$

To do this, we shall make use of the natural Banach algebra norm on $M_n(\mathbb{R})$, namely, the operator norm. So, in addition to the triangle inequality, there is also

$$\|AB\| \leq \|A\| \cdot \|B\| .$$

Lemma 9.4.4. *Let C be any matrix in $M_n(\mathbb{R})$ and $a > 0$ be an upper bound for all its entries. Then there exists $b > 0$ so that*

$$\|C^m\| \leq ba^m$$

for every positive integer m.

Proof.   Consider the geometric series $1 + u + u^2 + \cdots$ which converges when $|u| < 1$ to $\frac{1}{1-u}$ and apply this to $\frac{C_{i,j}}{a}$. Since by hypothesis $C_{i,j} < a$, this geometric series converges for every $i, j = 1, \ldots, n$. Hence, the series

$$I + \frac{C}{a} + \frac{C^2}{a^2} + \cdots$$

itself converges in $M_n(\mathbb{R})$. In particular, each of its n^2 numerical series converges. Taking b to be the maximum of these n^2 convergent sums provides our b, for which $\|C^m\| \leq ba^m$ for each m. $\square$

As we have just observed above, for each positive integer m, $\Phi(t + 2m\tau) = \Phi(t)C^m$. By compactness of $[0, 2\tau]$ and continuity of Φ, we know that Φ is bounded on $[0, 2\tau]$. Let $c > 0$ be a bound. Then for any $t \geq 0$, we write $t = 2m\tau + t_1$ for some integer m and t_1 in the interval. Hence, $\Phi(t) = \Phi(2m\tau + t_1) = \Phi(t_1)C^m$. Therefore,

$$\|\Phi(t)\| \leq a \|C^m\| \leq cba^m.$$

We want to find a $d > 0$ so that $|\Re(\lambda)| < d$ for all eigenvalues λ of C, but we must do it compatibly with the other choices already made. Now, since $a > |\lambda| > |\Re(\lambda)|$, all we have to do is take $d > a$. Then $|\Re(\lambda)| < d$ for all eigenvalues λ. Therefore, $\|C\|^m \leq e^{md}$ and using equation (9.4), we see that $\|\Phi(t)\| \leq cbe^{2\tau md}$. Since for t_1 in the interval $e^{t_1 d} \leq c_0$ for some c_0 and $e^{2\tau md} \leq c_1$, it follows that $\|\Phi(t)\| \leq cbc_0c_1e^{dt}$, and finally letting $R = cbc_0c_1$, we get $\|\Phi(t)\| \leq Re^{dt}$ for every $t \in [0, \infty)$.

It should be mentioned that periodic ODEs of the Floquet Liapunov type have played a role in more recent mathematics research where the author and Sacksteder proved in [11] that for any connected Lie group G with Lie algebra $\mathfrak{g}$, $G = \exp(\mathfrak{g}) \cdot \exp(\mathfrak{g})$. That is, any element of G is either an exponential or at most the product of two exponentials.

Chapter 10

Classical Partial Differential Equations of the Second Order

Now that we have studied the elements of ordinary differential equations (ODEs), we consider some elementary, but basic, partial differential equations. The usual abbreviation for *partial differential equation* is PDE. As mentioned in the introduction, these are equations where the unknown is a numerical function, say u, of several variables (here, we shall limit ourselves to two real variables) and the equation involves partial derivatives of u. So, a PDE in the dependent variable u and the independent variables x, y is an equation which is of the form

$$F(x, y, u, u_x, u_y, u_{xx}, u_{xy}, u_{yy}, u_{xxx}, u_{xxy}, \ldots) = 0,$$

where F is a function of the indicated quantities and at least one partial derivative occurs. As in ODEs, the *order* of a PDE means the order of the highest partial derivative occurring in the equation. The simplest example of a first-order PDE is $\frac{\partial u}{\partial y} = 0$, which, clearly, is satisfied by any function of the form $u(x, y) = f(x)$. Here, f is an arbitrary (smooth) function of x alone. The class of PDEs which will be of most interest is the class of second-order equations. We will be especially interested in certain *linear* second-order equations which often occur in physical applications. These are *linear second-order PDEs* with *constant coefficients* in u and independent variables $(x, y) \in \Omega$, where Ω is a domain in $\mathbb{R}^2$. They can be put in the form

$$au_{xx} + 2bu_{xy} + cu_{yy} + d_1 u_x + d_2 u_y + d_0 u = g(x, y), \qquad (10.1)$$

where a, b, c, d_1, d_2, d_0 are given real constants and at least one of a, b, c is different from zero, and $g(x, y)$ is a given smooth function defined on Ω. When $g \equiv 0$ on Ω, (10.1) is called *homogeneuous*.

A *solution* to (10.1) on $\Omega \subseteq \mathbb{R}^2$ is a function $u(x, y)$ of class $C^2(\Omega)$ which satisfies the PDE identically in Ω. The *general solution* of (10.1) is the set of all its solutions. To *solve* the PDE means to determine its general solution. As above, in a PDE, rather than the constant of integration that occurs in ODE, here there will be an arbitrary function of all the variables except the one with respect to which differentiation takes place.

In general, it may not be possible to write the general solution of a PDE in a closed form as we have done for many linear second-order ODEs. For this reason, it is important to have methods for combining known solutions when available. For homogeneous linear PDEs, we have the following simple rule. An easy consequence of the linearity of the PDE is the so-called *superposition principle*: If $u_1, \ldots, u_k$ are solutions of some homogeneous linear PDE and $c_1, \ldots, c_k$ are constants, then $c_1 u_1 + \cdots + c_n u_k$ is also a solution.

A fundamental technique for obtaining solutions of PDEs is the method of *separation of variables*. This means that we seek solutions of the form $u(x, y) = X(x)Y(y)$ and as a result obtain ODEs for $X(x)$ and $Y(y)$, which hopefully can be solved. The solution thus obtained is called a *separated solution*. When one independent variable represents *time*, we denote the function u by $u = u(x, t) = X(x)T(t)$. The methods of separation of variables and the superposition principle enable us to find solutions to boundary value problems and initial value problems for PDEs in the form of a convergent infinite series of variable separable solutions. This will be done in later sections of this chapter.

Some important examples of PDEs of the second order in two variables which we shall solve are the following:

- The *wave equation* $c^2 u_{xx} = u_{tt}$, or

$$c^2 \frac{\partial^2 u}{\partial x^2} = \frac{\partial^2 u}{\partial t^2},$$

where c^2 is a positive constant. This occurs in problems involving propagation of sound such as in a vibrating string. Here, $u(x, t)$ is the displacement of the string at position x and time t.

- *Laplace's equation* $u_{xx} + u_{yy} = 0$, or

$$\frac{\partial^2 u}{\partial x^2} + \frac{\partial^2 u}{\partial y^2} = 0.$$

This occurs in problems involving potentials and electrostatics. A solution to Laplace's equation is called a *harmonic* function.[1]

- The *heat equation* $c^2 u_{xx} = u_t$, or

$$c^2 \frac{\partial^2 u}{\partial x^2} = \frac{\partial u}{\partial t},$$

with c^2 a positive constant. This occurs in problems involving heat flow, such as in a metal rod. Here, $u(x, t)$ is the temperature of the rod at position x and time t.

All these PDEs have constant coefficients. Although we shall not deal with higher-dimensional analogs of these PDEs here, it is worthwhile mentioning these equations in more than two variables. These are as follows:

Let $x = (x_1, \ldots, x_n) \in \Omega \subseteq \mathbb{R}^n$. Here, one defines the Laplacian, Δ, in n variables by $\Delta u = \sum_{i=1}^{n} \frac{\partial^2 u}{\partial x_i^2}$. The *Laplace equation* is then

$$\Delta u = 0.$$

The *wave equation* is

$$c^2 \Delta u = \frac{\partial^2 u}{\partial t^2},$$

and the *heat equation* is

$$c^2 \Delta u = \frac{\partial u}{\partial t}.$$

In the last two equations, $x = (x_1, \ldots, x_n)$ are called the *space* variables, t is called the *time* variable, and in this case, $(x, t) = (x_1, \ldots, x_n, t) \in \mathbb{R}^{n+1}$. Note that in the *steady state*, i.e., when $\frac{\partial u}{\partial t} = 0$, these become Laplace's equation.

---

[1]Pierre Laplace (1749–1827) was a French mathematician who made fundamental contributions to the development of celestial mechanics, the theory of probability and mathematical physics.

10.1 Second-Order PDEs in Two Variables

In this section, we give a classification of the second-order linear PDE to canonical forms and find the general solution for each form. Then we will solve several problems involving the wave, Laplace's and heat equations. Finally, we will find *general solutions* of second-order linear homogeneous PDE with constant coefficients.

We begin with a historical example, namely, the wave equation which was first investigated and solved by both d'Alembert[2] and Euler in the mid-18th century. Here is d'Alembert's solution of the wave equation:

$$c^2 \frac{\partial^2 u}{\partial x^2} = \frac{\partial^2 u}{\partial t^2}. \tag{10.2}$$

First, note that, if φ and ψ are twice differentiable functions of one real variable, then a direct calculation using the chain rule shows that $u(x,t) = \varphi(x+ct) + \psi(x-ct)$ satisfies the wave equation. Proving the converse, is in effect, solving the wave equation. To do this, we perform a change of variables. Let $\xi = x + ct$ and $\eta = x - ct$ and see what the wave equation looks like in the new variables ξ and η. Solving for x and t, we get

$$x = \frac{\xi + \eta}{2} \quad \text{and} \quad t = \frac{\xi - \eta}{2c}.$$

Moreover, $\frac{\partial x}{\partial \xi} = \frac{\partial x}{\partial \eta} = \frac{1}{2}$, $\frac{\partial t}{\partial \xi} = \frac{1}{2c} = -\frac{\partial t}{\partial \eta}$ and $\frac{\partial^2 x}{\partial \eta \partial \xi} = 0 = \frac{\partial^2 t}{\partial \eta \partial \xi}$.
By the chain rule,

$$\frac{\partial u}{\partial \xi} = \frac{\partial u}{\partial x} \frac{\partial x}{\partial \xi} + \frac{\partial u}{\partial t} \frac{\partial t}{\partial \xi}.$$

Hence,

$$\frac{\partial^2 u}{\partial \eta \partial \xi} = \frac{\partial}{\partial \eta} \left(\frac{\partial u}{\partial x} \frac{\partial x}{\partial \xi} \right) + \frac{\partial}{\partial \eta} \left(\frac{\partial u}{\partial t} \frac{\partial t}{\partial \xi} \right)$$

---

[2]Jean-Baptiste d'Alembert (1717–1783) was a French mathematician, physicist and savant of the enlightenment. He was the leading mathematician of the group of *philosophes* who wrote the *Encyclopedie* (28 vols., 1751–1772) and was a friend of Euler, Laplace and Lagrange.

$$= \frac{\partial u}{\partial x}\frac{\partial^2 x}{\partial \eta \partial \xi} + \frac{\partial^2 u}{\partial x^2}\frac{\partial x}{\partial \eta}\frac{\partial x}{\partial \xi} + \frac{\partial u}{\partial t}\frac{\partial^2 t}{\partial \eta \partial \xi} + \frac{\partial^2 u}{\partial t^2}\frac{\partial t}{\partial \eta}\frac{\partial t}{\partial \xi}$$

$$= \frac{1}{4}\left[\frac{\partial^2 u}{\partial x^2} - \frac{1}{c^2}\frac{\partial^2 u}{\partial t^2}\right] = 0.$$

Thus, in the new variables, this equation becomes

$$\frac{\partial^2 u}{\partial \eta \partial \xi} = 0.$$

As $\frac{\partial^2 u}{\partial \eta \partial \xi} = \frac{\partial}{\partial \xi}\left(\frac{\partial u}{\partial \eta}\right) = 0$, we have to solve

$$\frac{\partial w}{\partial \xi} = 0 \quad \text{and} \quad \frac{\partial u}{\partial \eta} = w.$$

Now, $\frac{\partial w}{\partial \xi} = 0$ implies $w = f(\eta)$, where f is an arbitrary smooth function on η alone. The equation $\frac{\partial u}{\partial \eta} = w$, in turn, implies that

$$u(\xi, \eta) = \int_0^\eta f(s)ds + \varphi(\xi),$$

where φ is an arbitrary smooth function of ξ alone. But then,

$$u(\xi, \eta) = \psi(\eta) + \varphi(\xi),$$

where ψ is (again) an arbitrary smooth function with $\psi' = f$. Thus, returning to the old variables, we get

$$u(x, t) = \varphi(x + ct) + \psi(x - ct).$$

These solutions are known as *d'Alembert's solution*.

We now consider the general homogeneous linear second-order PDEs of the form (10.1). These have constant coefficients. For simplicity, we write

$$au_{xx} + 2bu_{xy} + cu_{yy} + \cdots = 0, \tag{10.3}$$

where $a, b, c \in \mathbb{R}$ with $a^2 + b^2 + c^2 \neq 0$ and the dots represent terms with derivatives of lower orders. This, of course, includes the wave equation, Laplace's equation and the heat equation. Equation (10.3) will be solved by methods quite analogous to those of ODEs, which were treated earlier.

The second-order terms are called the *principal part* (or the *symbol*) of the equation and, as we shall see, these will be decisive. The coefficient matrix of the principal part of the equation is the symmetric matrix,

$$A = \begin{bmatrix} a & b \\ b & c \end{bmatrix},$$

and the corresponding quadratic form of A is

$$Q(\xi, \eta) = a\xi^2 + 2b\xi\eta + c\eta.$$

The sign of the determinant $\det(A) = ac - b^2$ turns out to be invariant under smooth non-singular transformations of coordinates. In addition, it is possible to show that there exists a linear transformation of variables x, y, which reduces equation (10.3) to one of the following forms called *canonical forms*. (For this, the reader can consult pretty much any book on PDE.) In this respect, the role played by the sign of $\det(A)$ is decisive. Recalling Sylvester's criterion (Theorem 3.7.20) of [10], the sign of $\det(A)$ also determines the nature of the quadratic form Q.

1. If $\det(A) < 0$, i.e., $ac - b^2 < 0$, the equation is reducible by a linear change of variables to the form

$$u_{xx} - u_{yy} + \cdots = 0,$$

called the *hyperbolic form*. The quadratic form Q in this case is indefinite.

2. If $\det(A) > 0$, i.e., $ac - b^2 > 0$, the equation is reducible to the form

$$u_{xx} + u_{yy} + \cdots = 0,$$

called the *elliptic form*. Here, the quadratic form Q is strictly definite (positive or negative),

3. If $\det(A) = 0$, i.e., $ac - b^2 = 0$, the equation is reducible to the form

$$u_{xx} + \cdots = 0 \text{ or } u_{yy} + \cdots = 0,$$

called the *parabolic form*. Here, Q is degenerate.

In general, when the second-order linear non-homogeneous PDE involves more than two variables, it is of the form

$$\sum_{i,j=1}^{n} a_{ij} u_{x_i x_j} + \cdots = g(x),$$

and the classification to canonical forms is again based on the quadratic form $Q(v) = \langle Av, v \rangle$ associated with the coefficient (symmetric) matrix $A = (a_{ij})$. The equation is called *elliptic* if Q is strictly definite (positive or negative definite), *hyperbolic* if Q is indefinite but non-degenerate and *parabolic* if it is degenerate.[3] The main thing to keep in mind is that this classification and methods of solution for the three types of equations are often determinative.

Now, if we want to solve equation

$$a u_{xx} + 2b u_{xy} + c u_{yy} = 0, \tag{10.4}$$

for $u = u(x, y)$, consider the function $u(x, y) = \phi(y + \lambda x)$, where λ is an unknown constant to be determined later and ϕ an arbitrary (but *nonlinear*) smooth function of one variable. When does this u satisfy the equation?

A direct calculation using the chain rule shows $u(x, y) = \phi(y + \lambda x)$ satisfies the equation if and only if

$$\phi''(y + \lambda x)(a\lambda^2 + 2b\lambda + c) = 0.$$

Since this would have to hold for all x and y if the first factor were zero, this would mean the function ϕ is linear. Hence, the equation is satisfied for such a ϕ if and only if

$$a\lambda^2 + 2b\lambda + c = 0. \tag{10.5}$$

Equation (10.5) is called the *auxiliary equation* and of course can be solved by the quadratic formula. We may as well have assumed $a \neq 0$

---

[3] Actually, even when the coefficients of the PDE are *not* constant, $\sum_{i,j=1}^{n} a_{ij}(x) u_{x_i x_j} + \cdots = g(x)$ is said to be *elliptic, hyperbolic,* or *parabolic* in a domain Ω of $\mathbb{R}^n$ if it is, respectively, elliptic, hyperbolic or parabolic at every point x of Ω. (Things can get complicated because PDEs may be of different type in different parts of the region in which they are to be solved. A typical example of this is the Tricomi equation $y u_{xx} + u_{yy} = 0$.)

because if by chance $a = 0$, then we would merely reverse the roles of the variables so that c becomes a. The only problem here would be if $a = 0 = c$. Then our equation would be $bu_{xy} = 0$. Since if b is also 0, we have no equation at all, we can divide by b and get $u_{xy} = 0$, or

$$\frac{\partial^2 u}{\partial x \partial y} = 0,$$

which we have just solved. Its general solution is

$$u(x, y) = \varphi(x) + \psi(y),$$

where φ and ψ are arbitrary smooth functions in one variable.

Our objective is to reduce the general case to this one, where we are assured of having a second degree auxiliary equation (10.5). Now, as above in the case of ODEs, there are three possibilities depending on the *discriminant* $4(b^2 - ac)$ of the auxiliary equation:

1. $b^2 - ac > 0$, i.e., the roots are *real* and *distinct*: the *hyperbolic* case.
2. $b^2 - ac < 0$, i.e., the roots are *complex conjugates* of one another: the *elliptic* case.
3. $b^2 - ac = 0$, i.e., the roots are *real* and *equal*: the *parabolic* case.[4]

For example, the wave equation, $u_{xx} - c^2 u_{yy} = 0$, is hyperbolic since its auxiliary equation is $\lambda^2 - c^2 = 0$ with roots $\lambda_1 = c$ and $\lambda_2 = -c$. Laplace's equation, $u_{xx} + u_{yy} = 0$, is elliptic since its auxiliary equation is $\lambda^2 + 1 = 0$ with roots $\lambda_1 = i$ and $\lambda_2 = -i$, while for the heat equation, the principal part is $c^2 u_{xx} = 0$ and the equation is parabolic since its auxiliary equation is $c^2 \lambda^2 = 0$ and has equal roots, $\lambda_1 = \lambda_2 = 0$.

Actually, just as in ODE, we can deal with all three cases simultaneously. We solve the auxiliary equation for roots λ_1 and λ_2 which may be equal or not, or real or not. Then the auxiliary equation becomes $(\lambda - \lambda_1)(\lambda - \lambda_2) = 0$, and our equation becomes

$$(D - \lambda_1)(D - \lambda_2)u = 0,$$

---

[4]It is important to note that this classification agrees with the one given earlier in terms of the sign of the $\det(A) = ac - b^2$ since $b^2 - ac = -\det(A)$.

where $D = \frac{\partial}{\partial \xi}$ means differentiate with respect to $\xi = y + \lambda x$. Since these operators commute, it makes no difference which factor comes first. So, we have to solve $(D - \lambda_1)w = 0$, where $w = (D - \lambda_2)u$. Thus, just as in ODE, we have reduced our second-order equation to two simultaneous first-order equations,

$$(D - \lambda_1)w = 0$$

and

$$w = (D - \lambda_2)u.$$

1. **(Hyperbolic case):** If the roots are real and distinct, the equation $(D - \lambda_1)(D - \lambda_2)u = 0$ is exactly of the form $u_{\xi\eta} = 0$, where $\xi = y + \lambda_1 x$ and $\eta = y + \lambda_2 x$. Thus, as above, the general solution is

$$u = \varphi(y + \lambda_1 x) + \psi(y + \lambda_2 x),$$

where φ and ψ are arbitrary smooth functions of one variable.

2. **(Elliptic case):** The roots are complex conjugates. Here, just as in the case of distinct real roots, we get as the general solution

$$u = \varphi(y + \lambda_1 x) + \psi(y + \lambda_2 x),$$

where φ and ψ are arbitrary smooth functions of one variable. Only this time $\lambda_1 = \alpha + i\beta$ and $\lambda_2 = \alpha - i\beta$, where $\beta \neq 0$.

Let $\Phi = Re(\varphi) + iRe(\psi)$ and $\Psi = Re(\varphi) - iRe(\psi)$. It is easy to see, and we leave this as an exercise for the reader, that

$$u = u(x, y) = \varphi(y + \lambda_1 x) + \psi(y + \lambda_2 x)$$
$$= \Phi(y + (\alpha + i\beta)x) + \Phi(y + (\alpha - i\beta)x)$$
$$+ i(\Psi(y + (\alpha + i\beta)x) - \Psi(y + (\alpha - i\beta)x))$$

and that, by calculating the conjugate, the right-hand side is real. Thus,

$$u = \Phi(y + (\alpha + i\beta)x) + \Phi(y + (\alpha - i\beta)x) + i(\Psi(y + (\alpha + i\beta)x)$$
$$-\Psi(y + (\alpha - i\beta)x))$$

is the general solution in the *elliptic* case.

3. **(Parabolic case)**: If the roots are real and equal, we have

$$(D - \lambda_1)w = 0, \text{ where } w = (D - \lambda_1)u.$$

Hence, from the first equation: $w = f(y + \lambda_1 x)$ for some function f. So, we have to solve $(D - \lambda_1)(f(y + \lambda_1 x)) = 0$, and just as in the ODE case, we get the general solution

$$u = g(y + \lambda_1 x) + xf(y + \lambda_1 x),$$

where g and f are arbitrary smooth functions of one variable.

Due to its importance, we write explicitly the general solution for the *Laplace equation* $\Delta u = u_{xx} + u_{yy} = 0$ (this time without using holomorphic functions). As we observed, the roots of the auxiliary equation are $\pm i$. Thus, the general solution of the Laplace equation in two variables is

$$u(x, y) = \Phi(y + ix) + \Phi(y - ix) + i(\Psi(y + ix) - \Psi(y - ix)).$$

10.1.1   Boundary value problems for the wave equation

Next, we solve the wave equation satisfying *boundary* and *initial conditions* (such a problem is referred to as a *mixed-type problem*).

We first deal with small vibrations of a string. Here, we solve the wave equation,

$$c^2 u_{xx} = u_{tt}, \quad 0 < x < \pi, \ t > 0 \tag{10.6}$$

with the boundary conditions,

$$u(0, t) = u(\pi, t) = 0, \tag{10.7}$$

and the initial conditions,

$$u(x, 0) = f(x), \quad u_t(x, 0) = g(x), \tag{10.8}$$

where $f(x)$ and $g(x)$ are functions given in advance.

Solution. We first seek a (non-zero) solution in the form

$$u(x, t) = X(x)T(t), \tag{10.9}$$

satisfying the boundary conditions

$$u(0, t) = X(0)T(t) = 0 \quad \text{and} \quad u(\pi, t) = X(\pi)T(t) = 0 \text{ for all } t > 0.$$

Since we want a non-zero solution,

$$X(0) = X(\pi) = 0. \tag{10.10}$$

Substitution of the variable separable equation into the wave equation yields

$$c^2 X''(x)T(t) = X(x)T''(t)$$

or

$$\frac{X''(x)}{X(x)} = \frac{1}{c^2} \cdot \frac{T''(t)}{T(t)}.$$

But a function of x can be equal to a function of t only if both are equal to some constant real number, say $-\lambda$ (no assumption is made at this point as to whether λ is positive, negative or zero), i.e.,

$$\frac{X''(x)}{X(x)} = \frac{1}{c^2} \cdot \frac{T''(t)}{T(t)} = -\lambda.$$

This gives rise to two second-order linear ODEs, (10.11) and (10.12).

$$X'' + \lambda X = 0, \tag{10.11}$$
$$T'' + c^2 \lambda T = 0. \tag{10.12}$$

The differential equation (10.11) together with its boundary conditions is an eigenvalue problem which we noted can have non-trivial solution only if $\lambda > 0$ and the eigenvalues are $\lambda = \lambda_n = n^2$. The corresponding eigenfunctions (non-zero solutions) are

$$X(x) = X_n(x) = \sin(nx), \quad \text{for } n = 1, 2, 3, \ldots$$

(we have set $B = 1$). We now turn to equation (10.12). Replacing λ by its value $\lambda_n = n^2$, we get

$$T'' + c^2 n^2 T = 0.$$

Setting $\omega = nc$, this is the equation $T'' + \omega^2 T = 0$ we solved earlier. Its solution is $T(t) = A\cos(\omega t) + B\sin(\omega t)$, i.e.,

$$T(t) = T_n(t) = A_n \cos(nct) + B_n \sin(nct),$$

where A_n and B_n are arbitrary constants. We have constructed a set of solutions to the wave equation satisfying the prescribed boundary conditions given by

$$u_n(x,t) = X_n(x) \cdot T_n(t) = \sin(nx)[A_n \cos(nct) + B_n \sin(nct)]$$
$$(10.13)$$

for $n = 1, 2, 3, \ldots$ Now, by the superposition principle, any finite sum of these solutions

$$\sum_{n=1}^{N} u_n(x,t)$$

also satisfies our equation together with the boundary conditions. By continuity, the same is true for a convergent infinite sum of solutions:

$$u(x,t) = \sum_{n=1}^{\infty} \sin(nx)[A_n \cos(nct) + B_n \sin(nct)] \qquad (10.14)$$

if we can deal with the convergence problem, so that the series is twice differentiable termwise. For this, it will be necessary that the series be uniformly and absolutely convergent.

We now try to determine the constants A_n and B_n so that the infinite series in t satisfies the initial conditions. First,

$$u(x,0) = \sum_{n=1}^{\infty} A_n \sin(nx) = f(x)$$

must hold. Assuming termwise differentiation is valid, we must also have

$$\frac{\partial u}{\partial t}(x,0) = \sum_{n=1}^{\infty} cn B_n \sin(nx) = g(x).$$

Pursuing this further would take us into questions of *Fourier series*.[5],[6] By properties of differentiating convergent sums of such infinite "Fourier" series, any sum of these would also be a solution and, in fact, this process gives all the solutions. (There are a countable number of positive eigenvalues λ_n.) For the reader who is familiar with Fourier series, we note that assuming the functions $f(x)$ and $g(x)$ have a Fourier series expansion, we see that the A_n's must be the *Fourier coefficients* for $f(x)$, i.e.,

$$A_n = \frac{2}{\pi} \int_0^\pi f(x) \sin(nx) dx,$$

and the cnB_n must be the Fourier coefficients for $g(x)$, i.e.,

$$B_n = \frac{2}{cn\pi} \int_0^\pi g(x) \sin(nx) dx.$$

For more details concerning these issues, the reader is referred to Appendix D on real analytic functions and Appendix E on Fourier series.

We remark that the *principle of conservation of energy* holds here. The quantity

$$E(t) = \int_0^\pi \left[(u_t)^2 + (u_x)^2 \right] dx$$

is the *energy* of the solution for this problem. We will show it is constant. The first term of the integral denoted by $E_k(t) = \int_0^\pi (u_t)^2 dx$ (the integral of the square of the velocity of the vibrating string at the point x) represents the *kinetic energy*, while the second denoted

---

[5]By a *Fourier series* is meant a series of the form

$$\frac{1}{2}a_0 + \sum_{n=1}^\infty (a_n \cos nx + b_b \sin nx),$$

where a_0, a_n, b_n , $n = 1, 2, 3, \ldots$, are constants.

[6]Joseph Fourier (1768–1830) was a French mathematical physicist who accompanied Napoleon on his conquest of Egypt. He was a practical man as well, for upon his return to France when he became prefect of a district, he built its first roads.

by $E_p(t) = \int_0^\pi (u_x)^2 dx$ (the dilation of the string at the same point) represents the *potential energy* of the string. Now, the *total energy* $E(t) = E_k(t) + E_p(t)$ is *constant*. Indeed, its derivative,

$$\frac{dE}{dt} = \int_0^\pi \frac{\partial}{\partial t} \left[(u_t)^2 + (u_x)^2 \right] dx$$

$$= \int_0^\pi [2u_t u_{tt} + 2u_x u_{xt}] \, dx = 2c^2 \int_0^\pi [u_t u_{xx} + u_x u_{xt}] \, dx$$

$$= 2c^2 \int_0^\pi \left[\frac{\partial}{\partial x} (u_t u_x) \right] dx = 2c^2 u_t u_x \Big|_{x=0}^{x=\pi} = 0,$$

since $u_t(0, t) = u_t(\pi, t) = 0$.

10.1.2   Vibrations of an infinite string: d'Alembert's formula

The boundary conditions $u(0, t) = u(\pi, t) = 0$ in the above problem of a vibrating string tell us that the string remains motionless at the points $x = 0$ and $x = \pi$ (such points are called *nodes*). If a string is very long, then its ends will exert little influence on the vibrations occurring near the middle. Hence, when considering free vibrations of an infinite string, the problem reduces to the following initial value problem (also called the *Cauchy problem* for the two-dimensional wave equation).

Example 10.1.1. Find a twice continuously differentiable function $u(x, t)$ such that

$$c^2 u_{xx} = u_{tt}, \quad -\infty < x < \infty, \ t > 0, \tag{10.15}$$

satisfying the initial conditions

$$u(x, 0) = f(x), \quad u_t(x, 0) = g(x), \tag{10.16}$$

where $f \in C^2(\mathbb{R})$ and $g \in C^1(\mathbb{R})$ are given functions.

Solution. The general solution (d'Alemberts's solution) of the wave equation was found to be

$$u(x, t) = \varphi(x + ct) + \psi(x - ct). \tag{10.17}$$

Now, we use d'Alembert's solution to the initial value problem, i.e., to determine the functions φ and ψ. Using our initial conditions, we see that

$$\varphi(x) + \psi(x) = f(x) \quad \text{and} \quad c\phi'(x) - c\psi'(x) = g(x).$$

Differentiating the first equation and dividing by c in the second, we obtain

$$\varphi'(x) + \psi'(x) = f'(x),$$

$$\phi'(x) - \psi'(x) = \frac{1}{c}g(x).$$

This is a linear system of two equations in φ' and ψ'. The solution is

$$\varphi'(x) = \frac{1}{2}f'(x) + \frac{1}{2c}g(x), \quad \psi'(x) = \frac{1}{2}f'(x) - \frac{1}{2c}g(x),$$

which implies that

$$\varphi(x) = \frac{1}{2}f(x) + \frac{1}{2c}\int_0^x g(s)ds + C_1 \quad \text{and}$$

$$\psi(x) = \frac{1}{2}f(x) - \frac{1}{2c}\int_0^x g(s)ds + C_2.$$

Thus, the d'Alembert solution

$$u(x,t) = \varphi(x + ct) + \psi(x - ct)$$

yields

$$u(x,t) = \frac{1}{2}[f(x + ct) + f(x - ct)] + \frac{1}{2c}\int_{x-ct}^{x+ct} g(s)ds, \qquad (10.18)$$

where $C_1 + C_2 = 0$, since $u(x,0) = f(x)$. This is also known as d'Alembert's formula.

10.1.3   Boundary value problems for Laplace's equation

In this section, we consider two-dimensional boundary value problems for the Laplace equation. The typical boundary value problem for the Laplace equation is the *Dirichlet's problem.*[7]

Definition 10.1.2. *Dirichlet's problem.* Let Ω be a bounded domain (open connected set) in $\mathbb{R}^2$ with a piecewise smooth boundary $\partial(\Omega)$. If f is a given function which is defined and continuous on $\partial(\Omega)$, then the *Dirichlet problem* means finding a function $u \in C^2(\Omega)$ which is

1. defined and continuous on $\overline{\Omega} = \Omega \cup \partial(\Omega)$,
2. harmonic on Ω, i.e., $\Delta u = u_{xx} + u_{yy} = 0$ in Ω,
3. $u \equiv f$ on $\partial(\Omega)$.

The uniqueness of the solution was proved in our earlier work on ODE. The proof of existence of a solution of the Dirichlet problem in general is beyond the scope of this book. However, of particular importance are the cases when the Dirichlet problem can be solved *explicitly.* Here we solve the Dirichlet problem for a rectangle.

Example 10.1.3. Dirichlet's problem for a rectangle. Let $\Omega = \{(x, y) : 0 < x < a,\ 0 < y < b\}$ be a rectangle in $\mathbb{R}^2$. We solve the problem

$$\Delta u = 0 \text{ in } \Omega :, \tag{10.19}$$

$$u(0, y) = u(a, y) = 0, \tag{10.20}$$

$$u(x, 0) = 0, \quad u(x, b) = f(x). \tag{10.21}$$

Solution. We look for a variable separable (non-zero) solution,

$$u(x, y) = X(x)Y(y). \tag{10.22}$$

Substitution into the equation $\Delta u = 0$ and division by XY give

$$\frac{X''}{X} = -\frac{Y''}{Y}.$$

---

[7]Peter Dirichlet (1805–1859) made many contributions to analysis and number theory. In 1855, he succeeded Gauss as Professor at the University of Göttingen.

Reasoning as above, there must exist a constant λ such that

$$X'' + \lambda X = 0$$

for $0 < x < a$, and

$$Y'' - \lambda Y = 0$$

for $0 < y < b$. Moreover, the homogeneous boundary conditions give $X(0) = X(a) = 0$ and $Y(0) = 0$. The eigenvalue problem

$$X'' + \lambda X = 0, \quad X(0) = X(a) = 0$$

has eigenvalues $\lambda = \lambda_n = \left(\frac{n\pi}{a}\right)^2$, $n = 1, 2, 3, \ldots$, and eigenfunctions

$$X(x) = X_n(x) = \sin\left(\frac{n\pi x}{a}\right).$$

Turning to the variable y, setting $\beta_n^2 = \lambda_n$ and solving the ODE

$$Y'' - \beta_n^2 Y = 0,$$

we obtain

$$Y(y) = Y_n(y) = a_n e^{\beta_n y} + b_n e^{-\beta_n y}.$$

Here, it is preferable to get the general solution in terms of hyperbolic functions

$$Y_n(y) = A_n \cosh(\beta_n y) + B_n \sinh(\beta_n y).$$

Using the homogeneous boundary condition $Y(0) = 0$ (the non-homogeneous boundary condition for $y = b$ will be considered in the last step), we get $A_n = 0$ and obtain

$$Y_n(y) = B_n \sinh(\beta_n y) = B_n \sinh\left(\frac{n\pi y}{a}\right).$$

Thus, the functions

$$u_n(x, y) = X_n(x) \cdot Y_n(y) = B_n \sin\left(\frac{n\pi x}{a}\right) \sinh\left(\frac{n\pi y}{a}\right)$$

are solutions that satisfy the Laplace equation and all homogeneous boundary conditions. Superimposing (and assuming convergence), we arrive at the series solution

$$u(x, y) = \sum_{n=1}^{\infty} B_n \sin\left(\frac{n\pi x}{a}\right) \sinh\left(\frac{n\pi y}{a}\right).$$

This represents a harmonic function on the rectangle Ω and satisfies the homogeneous boundary conditions. Finally, we also require the solution to satisfy the boundary condition $u(x, b) = f(x)$, i.e.,

$$f(x) = u(x, b) = \sum_{n=1}^{\infty} B_n \sinh\left(\frac{n\pi b}{a}\right) \sin\left(\frac{n\pi x}{a}\right).$$

Setting $C_n = B_n \sinh\left(\frac{n\pi b}{a}\right)$, we see that

$$f(x) = \sum_{n=1}^{\infty} C_n \sin\left(\frac{n\pi x}{a}\right). \tag{10.23}$$

Pursuing this further would take us to questions of Fourier series. The reader who is familiar with the theory of Fourier series will recognize that foregoing expression for $f(x)$ is the Fourier series of the function f (assuming that f permits such an expansion) and so C_n is the Fourier coefficient

$$C_n = \frac{2}{a} \int_0^a f(x) \sin\left(\frac{n\pi x}{a}\right) dx.$$

In this way, we obtain formulas for the unknown coefficients B_n. In fact,

$$B_n = \frac{2}{a \sinh\left(\frac{n\pi b}{a}\right)} \int_0^a f(x) \sin\left(\frac{n\pi x}{a}\right) dx.$$

10.2   Laplace's Equation and Complex Analysis

In this section, we make the connection between harmonic functions in planar domains and functions of a complex variable. Suppose we have a *holomorphic* function $w = f(z)$ in a connected domain D in $\mathbb{R}^2$. Writing this complex-valued function as $f(z) = u(x, y) + iv(x, y)$ for $z = x + iy \in D$, because f is holomorphic, u and v satisfy the Cauchy–Riemann equations,

$$u_x = v_y, u_y = -v_x,$$

throughout D. Since u and v are smooth functions on D, the operators $\frac{\partial}{\partial x}$ and $\frac{\partial}{\partial y}$ commute. As a result, the reader should check that the Cauchy–Riemann equations imply that u and v are harmonic in D. They are called *conjugate* harmonic functions.

Now, let u merely be a solution to Laplace's equation, i.e., a harmonic function in two variables so that $\frac{\partial^2 u}{\partial x^2} + \frac{\partial^2 u}{\partial y^2} = 0$ in a connected domain D in $\mathbb{R}^2$. If D is simply connected as well, the converse of the above is also true, namely, u is the real part of a holomorphic function on D. This is because here (or on any manifold) connectivity is equivalent to D being arcwise connected so that any two points in D can be joined by a continuous contour in D. Now, the complex derivative of a holomorphic function satisfies $f'(z) = u_x - iu_y$. By the CR equations, since f is holomorphic, so is f'. Let z_0 be fixed in D and z be variable. Due to simple connectivity, by Cauchy's theorem, the contour integral

$$\int_{z_0}^z [u_x(\zeta) - iu_y(\zeta)]d\zeta$$

over any contour ζ from z_0 to z in D is independent of the path and depends only on the endpoint, z. Hence, it defines a holomorphic function of z on D whose real part is u. Thus, in this sense, we know what all the harmonic functions on D are, namely, they are the real parts of an arbitrary holomorphic function on D. This enables one to study harmonic functions of *two* real variables in a simply connected domain very effectively using complex analysis. For all this, see, e.g., [9]. In particular, concerning harmonic functions, a bounded harmonic function u on the entire plane must be constant. Another application is if u is harmonic and non-constant on a simply connected planar region, U, and D is a compact subdomain of U, then any maximum or minimum value of u on D must lie on the boundary, $\partial(D)$.

One final comment is in order. The statement of Laplace's equation alone doesn't imply that u is continuous and therefore doesn't imply u is the real part of a holomorphic function unless more is assumed about u. For example, let

$$u(x, y) = \frac{2xy}{(x^2 + y^2)^2}$$

if (x, y) is not the origin and $u(0.0) = 0$ otherwise. Thus, u is defined on the entire plane. Taking $x = y \neq 0$ and $x \to 0$, we see that $u(x, y) \to +\infty$, so u is spectacularly discontinuous at (0.0). On the other hand, differentiating does show u is harmonic on $\mathbb{R}^2 \setminus \{(0.0)\}$. Indeed, u is the real part of $f(z) = \frac{i}{z^2}$, which is holomorphic off $(0, 0)$, but cannot be extended holomorphically to the entire plane (since if it could, then u would certainly be continuous at the origin which it isn't). Finally, this example also shows the significance of the hypothesis of simple connectivity.

Exercise 10.2.1.

1. Show the real part of $\frac{i}{z^2}$ is $\frac{2xy}{(x^2+y^2)^2}$.
2. Check that the proof above shows that if u is a C^2 function on a simply connected domain, D, in $\mathbb{R}^2$ which is harmonic, then u is indeed the real part of a holomorphic function on D.

10.2.1   Boundary value problems for the heat equation

Finally, we solve the heat equation also by the separation of variables.

Example 10.2.2. Solve the heat equation

$$u_t - c^2 u_{xx} = 0, \quad 0 \leq x \leq L, \ t \geq 0,$$

satisfying the boundary conditions $u(0, t) = u(L, t) = 0$.

Solution. We are interested in (non-zero) variable separable solutions, $u(x, t) = X(x)T(t)$. These take the form

$$\frac{X''(x)}{X(x)} = c^{-2}\frac{T'(t)}{T(t)}.$$

As usual, this term is evidently independent of both x and t and so is constant $-\lambda$ and we get a system of two ODEs:

$$X''(x) + \lambda X(x) = 0 \quad \text{and} \quad T'(t) + \lambda c^2 T(t) = 0.$$

The boundary conditions give $X(0) = X(L) = 0$. The eigenvalue problem for the first equation has eigenvalues $\lambda_n = \frac{n^2\pi^2}{L^2}$ and eigenfunctions

$$X_n(x) = A_n \sin\left(\frac{n\pi x}{L}\right) \quad (n = 1, 2, \ldots),$$

where A_n are arbitrary constants. Substituting λ_n in the second (first-order) equation and solving the ODE, we find

$$T_n(t) = e^{-c^2 n^2 \pi^2 t/L^2}.$$

Thus,

$$u_n(x, t) = A_n e^{-c^2 n^2 \pi^2 t/L^2} \sin\left(\frac{n\pi x}{L}\right).$$

Superimposing solutions as before leads to a (convergent) series

$$u(x, t) = \sum_{n=1}^{\infty} A_n e^{-c^2 n^2 \pi^2 t/L^2} \sin\left(\frac{n\pi x}{L}\right),$$

which satisfies both the heat equation (provided the series $\sum_{n=1}^{\infty} |A_n| < \infty$ converges) and the boundary conditions $u(0, t) = u(L, t) = 0$.

If we also impose the initial condition $u(x, 0) = f(x)$, under the assumption that the function f has the Fourier series expansion, then

$$u(x, 0) = f(x) = \sum_{n=1}^{\infty} A_n \sin\left(\frac{n\pi x}{L}\right).$$

The numbers A_n can be determined by the formula:

$$A_n = \frac{2}{L} \int_0^L f(x) \sin\left(\frac{n\pi x}{L}\right) dx.$$

Finally, we consider the problem of the vibrating string. Here, we use Hamilton's principle of least action which is proved at the end of Chapter 11.

Consider a string which is clamped at both ends and then plucked (such as with a violin string). Here, we assume that the string is perfectly elastic (so, there is no potential energy stored in the string other than due to its height, i.e., to gravity) and that it is rigidly clamped at each end (as a violin string is). Let τ denote the tension in the string and L be its fixed equilibrium length at that tension. When plucked, the string vibrates in a vertical plane in such a way that each point on the string moves in a straight vertical line. We assume the amplitude of this vibration is so small that at every point and at all times, the slope of the tangent line to it is much less than 1. We also assume that the elongation is so slight as to have no effect on the tension τ. That is to say, τ is constant. We further assume that there is no loss due to friction with the air or anything else. Finally, we suppose uniform linear density, δ, of the string. Otherwise, instead of differential operators with constant coefficients, we would have to deal with the more difficult subject of differential operators with smooth, but variable coefficients.

We seek the shape of such a vibrating string which of course is a function of both position x and time t. Thus, we have two independent variables. Let $z = z(x,t)$ denote the height above the x-axis at time t, where $0 \le x \le L$ and $0 \le t$. Since the string is clamped at both ends, we know that $z(0,t) = 0 = z(L,t)$ for all t. Let dx be a small portion of the interval $[0,L]$ on the x-axis. The length of the corresponding piece of the string is $ds = \sqrt{1 + (\frac{\partial z}{\partial x})^2}dx$ and the corresponding potential energy due to stretching the string is τds. Thus, the total potential energy of the configuration is the difference between this and the initial or rest potential energy, τL.

Our assumption that $|\frac{\partial z}{\partial x}|$ being small throughout $[0,L]$ allows us to use the Taylor series of $\sqrt{1 + (\frac{\partial z}{\partial x})^2}$ to approximate this function it by its first two polynomial terms (in much the same way as we did when we studied the pendulum undergoing small vibrations). Thus, replacing $\sqrt{1 + (\frac{\partial z}{\partial x})^2}$ by $1 + \frac{1}{2}(\frac{\partial z}{\partial x})^2$ results in the total potential energy of the configuration taking the form

$$\frac{1}{2}\tau \int_0^L \left(\frac{\partial z}{\partial x}\right)^2 dx.$$

On the other hand, since δdx represents the mass of this same infinitely small piece of the string, the kinetic energy of this piece is $\frac{1}{2}\delta dx(\frac{\partial z}{\partial t})^2$ and so the total kinetic energy is $\int_0^L \frac{1}{2}\delta(\frac{\partial z}{\partial t})^2 dx$. Hence, for a particular time interval $[t_1, t_2]$, the *action* is given by

$$\frac{1}{2} \int_{t_1}^{t_2} \int_0^\delta \left(\frac{\partial z}{\partial t}\right)^2 dx dt - \tau \left(\frac{\partial z}{\partial x}\right)^2 dx dt.$$

By the *principle of least action*, we must minimize this double integral over the product space $[0, L] \times [t_1, t_2]$, subject to the boundary conditions on the edges of this rectangle, where $z(x, t) = 0$ for all $t \in [t_1, t_2]$ when $x \in [0, L]$ and in particular at t_1 and t_2 and $z(0, t) = z(L, t) = 0$ also in particular at t_1 and t_2. Thus, $z(x, t)$ is identically zero on the boundary. Accordingly, taking

$$F = \frac{1}{2} \left(\delta \left(\frac{\partial z}{\partial t}\right)^2 - \tau \left(\frac{\partial z}{\partial x}\right)^2\right),$$

we get for the Euler–Lagrange equation what is usually called the one (space)-dimensional wave equation. (These questions will be dealt with extensively in the following (and last) chapter of this book.)

$$\frac{\partial^2 z}{\partial x^2} = \frac{\delta}{\tau} \frac{\partial^2 z}{\partial t^2}.$$

Since the interval $[t_1, t_2]$ is arbitrary, this holds for all t and because the constant $\frac{\tau}{\delta} > 0$, one often calls this c^2. Thus, the *wave equation* takes the form,

$$c^2 \frac{\partial^2 z}{\partial x^2} = \frac{\partial^2 z}{\partial t^2}.$$

(These questions will be dealt with extensively in the following (and last) chapter of this book). To complete our analysis of the vibrating string, we only have to solve the wave equation, subject to the boundary conditions that $z(x, t)$ vanishes on the boundary, $[0, L] \times [0, a]$. We have already done this above. Of course, as a result, we will have solved any other physical problem satisfying the wave equation with boundary conditions. It might be worth mentioning that similar results and methods also hold for temperature distribution when heat is flowing in an insulated metal rod. By properties of differentiating convergent sums of such infinite "Fourier" series, any sum of these would also be a solution and, in fact, this process gives all the solutions. Although here the details are beyond the scope of this book, we can state the final result: There are a countable number of positive eigenvalues λ_n which are present in the most general solution to the one-dimensional space wave equation.

Chapter 11

An Introduction to the Calculus of Variations

In its simplest form, the calculus of variations is concerned with the following problem. Let $F(t, y, z)$ be a given smooth function of three real variables, where $a \leq t \leq b$ and y and $z \in \mathbb{R}$. Form the functional,

$$J(y) = \int_a^b F(t, y(t), y'(t))dt,$$

where $y(t)$ is a smooth function of t defined on $[a, b]$ and $y(a) = \alpha$ and $y(b) = \beta$ are given and fixed. If it exists, our objective is to find the function (or functions) $y(t)$ which minimize or maximize J subject to the boundary conditions, $y(a) = \alpha$ and $y(b) = \beta$. We shall see that, just as in the case of minimum and maximum problems in one real variable calculus, we will have necessary conditions and also sufficient conditions. The necessary condition in the case of numerical extrema is a numerical equation expressing the fact that, except for end points, at an extreme point, or even a local extreme point, the derivative must vanish, i.e., the point is a *critical* point. In the present situation the numerical function will be replaced by the functional J, points will be replaced by smooth curves in the domain of J and the numerical equation expressing the fact that the solution is a critical point will be replaced by a *differential equation* which expresses this vanishing, i.e., the extreme "point" is a critical point of J. This differential equation is called the Euler–Lagrange equation and will be of fundamental importance in what follows. After proving this sufficiency, we shall illustrate its use with a number of concrete problems from geometry, mechanics and geometric optics (this list could be increased). As we shall see, what has just been described is an extremely fruitful viewpoint. It affords a consistent methodology to deal with many diverse problems with very little technical difficulty (at least at the early stages!). As we move along in this subject from time to time, we shall introduce new problems whose solution will require various generalizations or extensions of the original formulation just posed, but whose core idea, however, will remain the same. These will be the exact analogs of going from one variable to several in the argument of F, but continuing to deal with curves. However, here, we will see that finding critical points will now involve solving *systems* of ODEs. Then we will turn to problems involving constraints, where we will employ an analog of Lagrange multipliers, or passing to problems in several *independent* variables (where the Euler–Lagrange equations in this case are PDEs), or combinations of either of these types with one or several constraints.

Theorem 11.0.1. *Let* $J(y) = \int_a^b F(t, y(t), y'(t))dt$ *(all as above). Suppose y is a minimum or maximum value of J. Then y satisfies the following differential equation called the Euler–Lagrange equation:*

$$\Gamma_y - \frac{d}{dt}\Gamma_{y'} = 0.$$

This will be proved in more general form in Theorem 11.2.2. Here is where the fundamental theorem of differential equations comes in (see Appendix B). Note that if $F_y - \frac{d}{dt}F_{y'} = 0$, then differentiating,

$$\frac{d}{dt}(F_{y'}(t, y, y')) = F_{y'} + F_{y'y}y' + F_{y'y'}y''.$$

Expressed in these terms, the Euler–Lagrange equation is

$$F_y = F_{y'} + F_{y'y}y' + F_{y'y'}y''.$$

If $F_{y'y'}$ is not identically zero, this is a linear, non-homogeneous equation of the second-order with variable, but smooth coefficients defined on a closed interval with fixed values at the end points. As such, if it has any solution, it has a *global* solution. (Of course, if the coefficient of y'' happens to be identically zero, then we merely have a first-order equation). Now, in this form, the uniqueness may be a problem. This occurs, for example, in the case of geodesics on a manifold of positive curvature where conjugate points occur. In particular, this occurs on the surface of the unit sphere in 3 space.

When F is independent of t, the situation has some special features and gives us the first integral of our second-order differential equation. In other words, we again have only to deal with a first-order equation because the second-order terms cancel. Our good fortune then allows us to solve it by quadratures.

Proposition 11.0.2. *Suppose $F = F(t, y, y')$ is independent of t and satisfies the Euler–Lagrange equation $F_y - \frac{d}{dt}F_{y'} = 0$. Then $F - y'F_{y'}$ is constant.*

Proof. We calculate $\frac{d}{dt}(F - y'F_z)$. This is $F_y y' + F_z y'' - \frac{d}{dt}(y'F_z)$. By the product rule, the last term is $y'\frac{d}{dt}F_z + y''F_z$. Thus, we get

$$F_y y' + F_z y'' - y'\frac{dF_z}{dt} - y''F_z.$$

Canceling second derivatives gives

$$F_y y' - y'\frac{dF_z}{dt} = y'\left(F_y - \frac{dF_z}{dt}\right) = 0$$

by the Euler–Lagrange equation. Hence, $F - y'F_z$ is constant. □

To get a feel for this, we now give some concrete examples of such problems and their solutions. We start with the simplest one.

11.0.1 The problem of shortest distance of a curve in the Euclidean plane

Let P and Q be fixed points in the plane (with different first coordinates) and consider all smooth curves $y = f(x)$ joining them. We wish to find the curve of shortest length connecting P and Q. Since the length l is expressed by $l = \int_a^b \sqrt{1 + y'^2}$, here $F(x, y, y') = \sqrt{1 + y'^2}$. Hence, not only is F independent of x, but it is also independent of y. Thus, $\sqrt{1 + y'^2} - \frac{y'^2}{1+y'^2} = c$ so that $1 = c(1 + y'^2)$. Evidently, $c \neq 0$. Hence, $1 + y'^2$, and therefore, y' is also constant. Thus, y is linear and can clearly be made to fit the initial data. Now, if by chance, the points had the same first coordinates, then we just perform a rotation of the plane. This preserves distance between points as well as lengths of curves and will create new points with different first coordinates. Rotating back shows that the shortest distance curve is always a straight line.

Later, we shall generalize this to geodesics on a Riemannian surface and solve the problem in the case of surfaces of constant curvature.

11.0.2   The Brachistochrone problem

Historically, this was the first open problem posed in the calculus of variations and was proposed by Giovanni Bernoulli[1] at the end of the 17th century. The problem is given two points P and Q lying in a vertical plane, with different first and second coordinates, find the path joining them of quickest descent. Let us suppose for definiteness that P is higher than Q and they are joined by a greased wire with a hollow bead of mass m placed at P (starting from rest, although this need not be so). The question is how should the wire be shaped so that the bead slides down to Q in the shortest possible time? For convenience, we can take coordinates $P = (0,0)$ and $Q = (a, b)$, where $a < 0$ and $b < 0$. Since $\frac{ds}{dt} = v$, where s is the arc length along the curve and v is the instantaneous velocity, we see that $dt = \frac{ds}{v}$ so that T, the total time of descent, is $T = \int_0^a \frac{ds}{v}$. Now, if the curve is given by $y = y(x)$, then $\frac{ds}{dx} = \sqrt{1 + y'^2}$. So, $ds = \sqrt{1 + y'^2}dx$. We calculate v by conservation of energy (thus the greased wire). Since at P the bead is at rest, the kinetic energy is zero. The potential energy is also zero at P since there the height is zero. Hence, $\frac{1}{2}mv^2 - mgy = 0$

---

[1]Giovanni Bernoulli (1667–1748) was one of a famous family of Swiss mathematicians. His brother Giacomo (1655–1705) was a collaborator and sometimes rival. They were the sons of a pharmacist who wanted one boy to study theology and the other medicine. Over his objections, both pursued careers in mathematics, making important discoveries in calculus, the calculus of variations, and differential equations. They sometimes worked together, but not without friction. Johann's (Giovanni) son Daniel (1700–1782) made important contributions to fluid dynamics (see Bernoulli's principle) and probability theory. Widely admired throughout Europe, he also studied and lectured on medicine, physics, astronomy, and botany.

so that $v = \sqrt{2gy}$. Since v and therefore the integral is independent of m, we see that the mass of the bead will be irrelevant to the question and

$$T = \frac{1}{\sqrt{2g}} \int_0^a \frac{\sqrt{1 + y'^2}}{\sqrt{y}} dx.$$

Thus, in the Brachistochrone (shortest time in Greek) problem, we can ignore $\frac{1}{\sqrt{2g}}$ and take

$$F(t, y, y') = \frac{\sqrt{1 + y'^2}}{\sqrt{y}}.$$

Since this is independent of t, applying the proposition above tells us that

$$\frac{y'^2}{\sqrt{y}\sqrt{1 + y'^2}} - \frac{\sqrt{1 + y'^2}}{\sqrt{y}} = c.$$

Multiplying this equation by $\sqrt{y}\sqrt{1 + y'^2}$, we see that $y'^2 - (1 + y'^2) = c\sqrt{y}\sqrt{1 + y'^2}$. (Here, $c \neq 0$ since otherwise this would say $1 = 0$.) That is, $\sqrt{y}\sqrt{1 + y'^2} = \frac{1}{c}$. This last constant is clearly positive, so it is $\sqrt{2\alpha}$ for some $\alpha > 0$. Hence, $y' = \sqrt{2\frac{\alpha}{y} - 1}$, and therefore, $x = \int \frac{\sqrt{y}dy}{\sqrt{2\alpha - y}}$. To calculate this integral, let $y = 2\alpha \sin^2 \frac{\theta}{2}$ and we get $x = 2\alpha \int \sin^2 \frac{\theta}{2} = \alpha(\theta - \sin(\theta)) + x_0$. But then $y = 2\alpha \sin^2 \frac{\theta}{2} = \alpha(1 - \cos(\theta))$. So, in parametric form, we have $x = \alpha(\theta - \sin \theta) + x_0$ and $y = \alpha(1 - \cos \theta)$. Since the curve passes through $(0, 0)$, we see $x_0 = 0$. These are the parametric equations of an upside down cycloid.

We now deal with the question of fitting the cycloid to initial data. We shall show that α can be adjusted to have the cycloid pass through $Q = (a, b)$ in the first loop and that this uniquely determines the cycloid. This gives an absolute minimum time of descent which can be calculated from the integral expressing T.

Lemma 11.0.3. *If $x = \alpha(t - \sin t)$ and $y = \alpha(1 - \cos t)$ be the parametric equations of a cycloid and $f(t) = \frac{t - \sin t}{1 - \cos t}$, then this cycloid passes through (a, b) if and only if for some $t_0 \in [0, 2\pi)$, $f(t_0) = \frac{a}{b}$.*

Proof. If (a, b) lies on the cycloid in the first loop, then this is clear. Conversely, suppose $f(t_0) = \frac{a}{b}$. Then $\frac{b}{1-\cos t_0} = \frac{a}{t_0 - \sin t_0}$. Let α be this common value. Then $x(t_0) = \alpha(t_0 - \sin t_0) = a$ and $y(t_0) = \alpha(1 - \cos t_0) = b$.   □

Using this lemma, we show that given any a and b both negative, there is a unique cycloid passing through (a, b) in the first loop. We calculate $\lim_{t \to 0^+} f(t)$. By L'Hospital's rule, differentiating numerator and denominator twice, we see that this limit is zero. Hence, by continuity, $f(0) = 0$. On the other hand, $\lim_{t \uparrow 2\pi} f(t)$ is clearly $\frac{2\pi}{0^+} = +\infty$. Since $\frac{a}{b} > 0$, it follows from the intermediate value theorem that for some $t_0 \in [0, 2\pi)$, $f(t_0) = \frac{a}{b}$. Clearly, this can then be done in $[0, \pi)$. We remark that Bernoulli solved this problem, but by a different method, then the above. The solution we have presented here is due to Euler. In a later section, we shall discuss Bernoulli's original solution and its significance.

11.0.3   The isoperimetric problem

We now turn to the so-called *isoperimetric problem*. That is, consider all smooth simple closed planar curves $(x(t), y(t))$ of a given (fixed) length, say, l. The problem is to find the curve which encloses the largest area and find the relationship of this largest area, A, to the length, l.

Suppose we had a curve which gave the largest area. Choose two points on the curve and in this way divide the curve into two subcurves. Clearly, if the original curve were to enclose the largest area, each of these subcurves would have to be convex. In particular, the line segment joining the two points would intersect the curve nowhere else. Now, consider these points, as above, but with the additional requirement that the length along the curve joining them (in either direction) is $\frac{l}{2}$. Then the areas enclosed must be equal, for otherwise this would clearly violate the maximal area property which we are assuming. Thus, we are reduced to considering the interval, say, $(0, 0)$ to $(a, 0)$ on the real axis and a convex curve defined on this interval lying in the first quadrant of length $\frac{l}{2}$ and enclosing an area, say A. We may also assume the curve $(x(s), y(s))$ is parametrized by arc length s. Now, since $A = \int_0^a y \, dx$, we see that $A = \int_0^{\frac{l}{2}} y(s) x'(s) \, ds$, where $y(0) = 0 = y(\frac{l}{2})$. Since we have chosen arc length as parameter, we know $x'^2(s) + y'^2(s) \equiv 1$ for all s. Here, of course, A depends on the curve (i.e., in this case depends only on y), so we can write

$$A(y) = \int_0^{\frac{l}{2}} y(s)\sqrt{1 - y'^2(s)}\,ds,$$

where $y(0) = 0 = y(\frac{l}{2})$ and our task is to see what the condition that $A(y)$ be maximal forces on y.

Here, $F(s, y, y') = y\sqrt{1 - y'^2}$ and since the integrand is independent of s, Proposition 11.0.2 tells us that there is a constant, c where $-y = c\sqrt{1 - y'^2}$. If $c = 0$, then evidently $y = 0$ and the area enclosed is zero. This is the minimal area solution which doesn't interest us at all. So, we may assume $c \neq 0$. Dividing by c and solving for y', we get $y' = \frac{\sqrt{c^2 - y^2}}{c}$. We can solve this first-order differential equation by separation of variables and the substitution $y = c\sin(\theta)$ and get $y(s) = c\sin\frac{s+\alpha}{c}$. Using the boundary conditions and the fact that the curve is convex, we see that $\alpha = 0$ and $\frac{l}{2c} = \pi$. Therefore,

$$y(s) = \frac{l}{2\pi}\sin\frac{2\pi s}{l}.$$

Now, since $x'^2(s) + y'^2(s) \equiv 1$, we can then also solve for $x(s)$ using the fact that $x(0) = 0$ and get

$$x(s) = \frac{l}{2\pi}\left(1 - \cos\left(\frac{2\pi s}{l}\right)\right).$$

Thus, $(x(s) - \frac{l}{2\pi})^2 + y^2(s) \equiv \frac{l^2}{4\pi^2}$. This means that our subcurve lies on a circle centered at $(\frac{l}{2\pi}, 0)$ and of radius $\frac{l}{2\pi}$ so that the area enclosed by the original curve is $\pi(\frac{l}{2\pi})^2 = \frac{l^2}{4\pi}$, thus proving the solution to the isoperimetric problem is a circle and also proving the isoperimetric inequality, A, the enclosed area, is $\leq \frac{l^2}{4\pi}$, with equality holding if and only if the curve is a circle.

We remark that the isoperimetric problem can also be solved using Wirtinger's inequality and Fourier series. Also, it clearly could be considered a variational problem with a constraint, namely, the arc length has constant value l. We shall also solve it on that basis in what follows and compare the result with what we have learned here.

We also remark that the isoperimetric inequality can be extended to $\mathbb{R}^n$, where $n \geq 3$ as follows: If Ω is a bounded domain in $\mathbb{R}^n$ with smooth boundary, $\partial\Omega$, and the n and $n-1$ volumes are ν_n and ν_{n-1}, respectively, then

$$[\nu_n(\Omega)]^{n-1} \leq \frac{[\nu_{n-1}(\partial\Omega)^n]}{n^n c_n},$$

where c_n is the volume of the unit sphere in $\mathbb{R}^n$. The reader should check that this agrees with what we just found when $n = 2$.

11.0.4   The minimal surface problem for a surface of revolution

Let P and Q be fixed points in the plane with different first coordinates and consider all smooth curves $y = f(x)$ joining them. The problem is to find a curve which minimizes the area of the resulting surface of revolution about the x-axis. Here, $A(y) = \int_a^b 2\pi y\sqrt{1+y'^2}dx$. Hence, $F(x,y,y') = y\sqrt{1+y'^2}$. Since F is independent of x, we know, that if there were a solution, $F - y'F_{y'}$ would be constant. Thus, $c\sqrt{1+y'^2} = -y$. Squaring, we get $c^2(1+y'^2) = y^2$. Clearly, $c \neq 0$ since then $y = 0$ and would not fit the initial data. Therefore, $y' = \frac{\sqrt{y^2-c^2}}{c}$. Separating variables and integrating give $c\int \frac{dy}{\sqrt{y^2-c^2}} = x - k$. To calculate this integral, let $y = c\cosh t$. Then $y(x) = c\cosh(\frac{x-k}{c})$, a catenary. We leave it to the reader to show that this can be uniquely made to fit the initial data.

After we deal with problems involving several independent variables in the following section, we will be able to drop the requirement that we have a surface of revolution.

11.0.5   The minimal surface problem

We now give an example where the minimal *volume* problem for a surface of revolution has no solution. Finally, although it seems rather similar to the above minimal surface area problem, here we actually have an example of a variational problem with no solution. Let P and Q be fixed points of the plane with different first coordinates and consider all smooth curves $y = f(x)$ joining them. We now try to find a curve which minimizes the *volume* of the resulting surface of revolution about the x-axis. Here, $V(y) = \pi \int_a^b y^2 dx$, and hence, $F(t, y, y') = y^2$. Since F is independent of t, we know that if there were a solution, $F - y'F_{y'}$ is constant. However, since here $F_{y'} = 0$, we see that F is constant. But $F = y^2$. Thus, y is constant. But there is no constant passing through P and Q since they have different second coordinates, a contradiction. Actually, what goes wrong here is more fundamental. For even if the heights were equal, our calculation shows that if there were a solution, it would have to be the constant function. But this clearly doesn't minimize the volume of the revolved figure since we can get smooth curves y_n passing through the end points which are on the x-axis for most of the curve. Hence, $\inf V(y_n) = 0$. Therefore, there is no minimum, since the zero function doesn't pass through the end points. Here, there is an infimum of the volumes of all these curves, namely, zero. *But it is not achieved by any curve satisfying the required boundary conditions.*

We conclude this section with a (continuous) generalization of Snell's law which states that when light passes through a boundary between two different isotropic media, such as water or glass, the angle of incidence $\theta(A)$ is related to the angle of refraction $\theta(B)$ by

$$\frac{\sin(\theta(A))}{\nu_A} = \frac{\sin(\theta(B))}{\nu_B},$$

where ν_A and ν_B are the respective indices of refraction of A and B.

11.1 The Problem of Geometric Optics

Given two points P and Q lying in a plane in which the optical density varies smoothly from point to point, our problem here is to find the path joining them along which light takes the least time to travel. Since just as in the brachistochrone problem $\frac{ds}{dt} = v$, where s is the arc length along the curve and v is the instantaneous velocity, we see that $dt = \frac{ds}{v}$ so that T, the total time, is given by $T = \int_0^a \frac{ds}{v}$. Now, if the curve lying in this plane is $y = y(x)$, then $\frac{ds}{dx} = \sqrt{1 + y'^2}$. So, $ds = \sqrt{1 + y'^2} dx$, and hence, $T = \int_a^b \frac{\sqrt{1+y'^2}}{v(x,y)} dx$, where $v(x, y)$ is the velocity of light in the medium at the point (x, y) in the plane.

Now, the reader will note that this is very similar to the brachistochrone problem, except here v is an unknown function of both x
and y instead of a known function of y alone. In any case, the integrand of the functional we have to minimize is $F = \frac{\sqrt{1+y'^2}}{v(x,y)}$.[2] If v is
independent of x, then so is F and we could apply our usual methods.
We can now explain how Giovanni Bernoulli understood and solved
the brachistochrone problem at the end of the 17th century before
the calculus of variations was invented!

We first consider a related extremalization problem from calculus
of one variable. Consider two materials of different uniform optical
densities in the form of contiguous strips, A and B, where v_A and
v_B are the respective velocities of light in A and B. Let $P = (a, a')$
be a point of A and $Q = (b, b')$ be a point of B. Since in each media
separately the velocity of light is constant, the problem of shortest
time path has the same solution as the problem of shortest distance
path. Thus, the solution is a geodesic of the plane which we have
already seen is a straight line. Since within A and B light travels in
straight lines, if a path were to join P and Q, the only possibility
would be for it to change angle at the interface of A and B. Choose
coordinates so the interface point on the trajectory is $(x, 0)$, where x
is the unknown. Then the total time of the trip is

$$T(x) = \frac{\sqrt{a'^2 + (a-x)^2}}{v_A} + \frac{\sqrt{b'^2 + (b-x)^2}}{v_B}.$$

Calculating the derivative and setting this equal to zero, we get

$$\frac{x-a}{v_A\sqrt{a'^2 + (a-x)^2}} = \frac{x-b}{v_B\sqrt{b'^2 + (b-x)^2}}.$$

That is to say,

$$\frac{\sin(\theta_A)}{v_A} = \frac{\sin(\theta_B)}{v_B},$$

where θ_A and θ_B are the respective angles of incidence with the
normal to the boundary at $(x, 0)$. This is Snell's law. The quickest

---

[2]As the reader will see, we will reencounter this situation when we investigate
the geodesics in the upper half plane.

such path is the one where the ratio of sines of angles is equal to the ratios of the velocities. It says that as light enters a more dense optical medium and slows down, it bends more toward the normal to make up for this. Knowing this fact, Bernoulli considered a non-homogeneous medium, but one in which the optical density depended only on the vertical coordinate y, where perhaps one doesn't know the exact nature of the dependence, only that it is smooth. He then divided this media (plane) into an infinite number of very thin horizontal strips of thickness dy. These strips being so thin that the velocity of light within each one was essentially constant, so that as the light passed from one of these strips to the next, it would behave exactly as above and would obey Snell's law. Therefore, at the boundary, we would have for all y, i.e., for all strips, $\frac{\sin(\theta(y))}{v(y)} = c$, a constant. However, since θ is the angle with (each) normal if we let ϕ be the complementary angle, i.e., the angle with the x-axis, we see that $\sin(\theta(y)) = cv(y)$ and since this is the angle whose tangent is the derivative $\frac{dy}{dx}$, we see that $\cot^2(\theta(y)) = (\frac{dy}{dx})^2$. Since $\csc^2(\theta(y)) = \cot^2(\theta(y)) + 1$, we get $\sin(\theta(y)) = \frac{1}{1+(\frac{dy}{dx})^2}$. Thus,

$$cv(y) = \frac{1}{1 + (\frac{dy}{dx})^2},$$

and solving for $\frac{dy}{dx}$, we have a simple first-order equation

$$y' = \pm \frac{\sqrt{1 - c^2 v(y)^2}}{cv(y)},$$

in which the variables separate. Thus,

$$x = \pm c \int \frac{v(y)dy}{\sqrt{1 - c^2 v(y)^2}} + c',$$

the constants c and c' being determined by P and Q. When $v(y) = \sqrt{y}$, this is the brachistochrone problem. Otherwise, it is a generalization and an important one. We leave to the reader as an exercise to show that if one considers this as a variational problem, then the solution is the same as Bernoulli's 400 years ago. Thus, we have also solved the following problem of geometric optics: Given a smooth function $v(y)$, find the planer path joining the points P and Q

which minimizes the functional $J(y) = \int \frac{\sqrt{1+y'^2}}{v(y)}dx$. The solution is the equation above. For example, as a consequence, we see that when solar light enters the Earth's atmosphere obliquely (to the normal), since the optical density increases smoothly, the light bends smoothly more and more toward the normal.

11.1.1   Sufficient conditions

In 1,2,4, and 6, we have to prove that our solution is actually a *minimum*. This means that we must look at sufficient conditions for a minimum to occur. What is available to us is Corollary 11.2.5. In this situation, it says that if F is independent of y and $F_{y',y'} \geq 0$, then each critical point is actually a global minimum. Now, in cases 1 and 4, these conditions are easily verified and are left to the reader. However, in cases 2 and 6, $F = p(x,y)\sqrt{1+y'^2}$, where p is a positive smooth function and so although $F_{y',y'} = \frac{p(x,y)}{(1+y'^2)^{\frac{3}{2}}} \geq 0$, our criterion is not sufficient since here F depends on y. Similarly in 3, Corollary 11.2.5 is not good enough to see that we actually get a maximum since although here $F_{y',y'}$ is positive, F again depends on y. What is needed to deal with this situation and be sure that there is a solution to the extremal problem being considered are methods of functional analysis, especially compactness in certain function spaces (and as such is beyond the scope of this book). This would guarantee that the sup or inf is actually achieved. Once we know that this is so and therefore that a solution exists, we can use the Euler–Lagrange equations to find it. In this sense, here also, the situation very much resembles that of calculus in one or several real variables where compactness of the domain is what guarantees the existence of a solution, which is then found by sifting through the critical points (and boundary points!) (see [5]).

11.2   Constraints

11.2.1   The catenary

We now consider the problem of a hanging flexible cable of fixed length between two points. The physical principle here is that it hangs in such a way that its potential energy is minimal. However, we also have a constraint, namely, the length is fixed. Thus, we have two functionals involved, the potential energy, $J(y) = \int_a^b F(x, y, y')dx$, and the length, $J_0(y) = \int_a^b F_0(x, y, y')dx = c$. We want to extremize J subject to the condition that J_0 is constant, so $J_0 = c$. As we shall see, the way to do this is to form $J^* = J + \lambda J_0$, where λ is a real parameter (Lagrange multiplier) to be determined. Then J^* is the functional associated with $F^* = F + \lambda F_0$. By Theorem 11.3.1, which deals with these questions in a somewhat more general form, any extreme value $y = y(x)$ of J taking prescribed values at the end points which also takes the fixed value c when we apply J_0 must satisfy the Euler–Lagrange equation for F^*, namely, $F_y^* - \frac{d}{dx}F_{y'}^* = 0$.

In the present situation, let l be the length of the cable and δ be the constant linear density. Thus, an infinitely small piece of the cable of length ds, where s is the arc length has weight $g\delta ds$ and therefore potential energy $yg\delta ds$. Thus, $V(y) = g\delta \int_0^l y\sqrt{1 + y'^2}dx$ which we want to minimize while holding $\int_0^l \sqrt{1 + y'^2}dx$ constant. Hence,

$$F^* = g\delta y\sqrt{1 + y'^2} - \lambda\sqrt{1 + y'^2}.$$

Now, F^* is independent of x. Applying Theorem 11.3.1, we see $F^* - y'F_{y'}^*$ is constant so that

$$(g\delta y + \lambda)\left(\frac{y'^2}{\sqrt{1 + y'^2}} - \sqrt{1 + y'^2}\right) = c.$$

From this, we get, by separation of variables as in problem 4,

$$y = \frac{\lambda}{g\delta} - \frac{c}{g\delta}\cosh\left(\frac{g\delta(x - a)}{c}\right),$$

where a is a constant of integration. The three constants, λ, c and a, can then be used to fit the solution to the initial data. Note that we have chosen the left end point to have height zero in calculating the potential energy. Therefore, there are exactly three parameters to determine.

Another example of an extremal problem with a constraint is provided by the isoperimetric problem we solved earlier. For instructional purposes, here, we resolve it by Lagrange multipliers.

We have $\int_a^b \sqrt{1 + y'^2}\,dx = l$, the fixed length, while we want to maximize $A = \int_a^b y\,dx$, the area. Hence, $F(x, y, z) = y$ and $G(x, y, z) = \sqrt{1 + z^2}$. Therefore, $(F - \lambda G)(x, y, z) = y - \lambda\sqrt{1 + z^2}$, where λ is fixed. Since this function must satisfy the Euler–Lagrange equation by Theorem 11.3.1 and here $F - \lambda G$ is independent of x, we get $z(F - \lambda G)_z - (F - \lambda G)$ is constant. Writing this out tells us that

$$\frac{-\lambda z^2}{\sqrt{1 + z^2}} - y + \lambda\sqrt{1 + z^2} = c.$$

This means that $\lambda = (y + c)\sqrt{1 + y'^2}$. Calling $y + c = u$ so that $y' = u'$, we get $\lambda = u\sqrt{1 + u'^2}$. We solve for u' and get $\frac{du}{dx} = \frac{\sqrt{\lambda^2 - u^2}}{u}$. We now separate the variables $\int dx = \int \frac{u\,du}{\sqrt{\lambda^2 - u^2}}$. Integrating and squaring tells us that $(x + \alpha)^2 + u^2 = \lambda^2$, where α is a constant of integration. Thus, in terms of the original parameters, we have a circle $(x + \alpha)^2 + (y + c)^2 = \lambda^2$ of radius, the Lagrange multiplier, and so λ is determined by the initial information, namely, the arc length of the curve.

11.2.2   Vector-valued curves

We now turn to the situation where our functional J is more elaborate and we will illustrate why this generalization is necessary with some significant examples. Let n be an integer and $F(t, Y, Z)$ be a given smooth function of $2n + 1$ real variables where $a \le t \le b$ and Y and $Z \in \mathbb{R}^n$. Form $J(Y) = \int_a^b F(t, Y(t), Y'(t))\,dt$, where $Y(t)$ is a smooth vector-valued curve parametrized by $t \in [a, b]$, with $Y'(t)$ its derivative and $Y(a) = \alpha$ and $Y(b) = \beta$ are given and fixed in $\mathbb{R}^n$. If it exists, our present objective is to find the curve $Y(t)$ in $\mathbb{R}^n$ which extremizes J subject to the boundary conditions, $y(a) = \alpha$ and $y(b) = \beta$. In this context, our original formulation was the case when $n = 1$. Here, we shall see that, just as above, we have necessary conditions and also sufficient conditions. However, now, the necessary condition expressing the fact that an extreme point is a critical point will not be a numerical differential equation, but rather a *system* of n differential equations, also called the Euler–Lagrange equations (see Theorem 11.2.2). For the sufficient condition, here instead of the second derivative $F_{y',y'}$ being ≥ 0 or ≤ 0, now the sufficient condition will state that the $n \times n$ symmetric matrix of second derivatives is positive semidefinite or negative semidefinite (see Corollary 11.2.5). The reader will note that, in addition to generalizing the calculus of variations in one variable, these results are exact analogs of the usual corresponding ones to those for minimum and maximum problems in calculus of several real variables.

We denote by $\langle , \rangle$ the standard inner product on $\mathbb{R}^n$. Here is the fundamental lemma in the more general form that is needed. Its proof will devolve to the case $n = 1$ which we also prove in the following. In this way, we will have filled in the missing result which was needed to complete the proofs when $n = 1$.

Lemma 11.2.1. *Let f be a continuous function $f : [a, b] \to \mathbb{R}^n$ and suppose $\int_a^b \langle f(t), g(t) \rangle dt = 0$ for all smooth functions $g : [a, b] \to \mathbb{R}^n$, satisfying $g(a) = 0 = g(b)$. Then $f \equiv 0$ on $[a, b]$.*

Proof. Since f is continuous and (a, b) is dense in $[a, b]$, it suffices to prove that $f \equiv 0$ on (a, b). Suppose contrarywise that $f(t_0) \neq 0$ for some $t_0 \in (a, b)$. Since $f(t_0) \neq 0$, it must be non-zero in some coordinate, say $*$. Assuming the lemma to be true for $n = 1$, we can find a numerical function g_* with $g_*(a) = 0 = g_*(b)$ and $\int_a^b f(t)g_*(t)dt \neq 0$. Take $g(t)$ to be zero in all coordinates except $*$. In that coordinate, we take the value to be $g_*(t)$ for all $t \in [a, b]$. Then $g(a) = 0 = g(b)$ and $\int_a^b \langle f(t), g(t) \rangle dt \neq 0$. This is a contradiction.

Now, when $n = 1$, we again have $f \neq 0$ on some small open interval I about $t_0 \in (a, b)$, where $|f(t)| \geq \frac{1}{2}|f(t_0)|$ everywhere on I. By Urysohn's Lemma in Appendix C to get test functions, choose a smooth positive function g supported on I. Then $\int_a^b f(t)g(t)dt = \int_{I-} f(t)g(t)dt$. But the latter is clearly non-zero, a contradiction. $\square$

As a consequence, we get the following theorem.

Theorem 11.2.2. *Let $J(Y) = \int_a^b F(t, Y(t), Y'(t))dt$ and suppose Y is an extreme value of J. Then it satisfies the following system of differential equations for $i = 1, \ldots, n$:*

$$F_{y_i} - \frac{d}{dt} F'_{y_i} = 0,$$

where $Y(t) = (y_1(t), \ldots, y_n(t))$ and $Y'(t) = (y_1'(t), \ldots, y_n'(t))$. Written in vector form, we have $F_Y - \frac{d}{dt} F_Z = 0$.

Proof.   Let η be an arbitrary smooth function, $\eta : [a, b] \to \mathbb{R}^n$, satisfying $\eta(a) = 0 = \eta(b)$ and $\epsilon > 0$. Consider the function $Y(t) + s\eta(t)$ on $[a, b]$, where $|s| < \epsilon$. For each such s, this function is smooth, vector-valued and takes the required values at the end points a and b. Assuming that we are in an open set of functions by taking ϵ small enough, $Y(t) + s\eta(t)$ is in this set for all s if $Y(t)$ itself is. Hence, in the case of a minimizer, $J(Y + s\eta) \geq J(Y)$ for all $|s| < \epsilon$. Therefore, by calculus of one real variable, $\frac{d}{ds} J(Y + s\eta)|_{s=0} = 0$. The proof for a maximizer is completely similar. Now, by differentiation under the integral sign, $\frac{d}{ds} J(Y + s\eta) = \int_a^b \frac{d}{ds} F(t, Y^*(t), Y^{*\prime}(t)) dt$, where $Y^* = Y + s\eta$ and $Y^{*\prime}$ is its derivative with respect to t. Consider the first-order Taylor expansion of F at the point (t, Y, Z).

$$F(t, Y^*, Z^*) = F(t, Y, Z) + s(F_{y_1} \eta_1 + \cdots + F_{y_n} \eta_n$$
$$+ F_{z_1} \eta_1' + \cdots + F_{z_n} \eta_n') + O(s^2).$$

In terms of the inner product on $\mathbb{R}^n$, we have

$$F(t, Y^*, Z^*) = F(t, Y, Z) + s(\langle F_Y, \eta \rangle) + \langle F_Z, \eta' \rangle + O(s^2).$$

Differentiating with respect to s and integrating over $[a, b]$ with respect to t gives

$$\frac{d}{ds} J(Y^*) = \int_a^b (\langle F_Y, \eta \rangle) + \langle F_Z, \eta' \rangle) + O(s) dt.$$

Then using the linearity of the integral and taking the limit as $s \to 0$, it follows that

$$\int_a^b \left(\left\langle \frac{\partial F}{\partial Y}, \eta \right\rangle + \left\langle \frac{\partial F}{\partial Z}, \eta' \right\rangle \right) dt = \frac{d}{ds} J(Y + s\eta)|_{s=0} = 0.$$

Calculating each of the individual terms of the η' part via integration by parts, we get

$$\left\langle \frac{\partial F}{\partial Z}, \eta \right\rangle \Big|_a^b + \int_a^b \left\langle -\frac{d}{dt} F_Z, \eta \right\rangle dt.$$

Taking into account the boundary values of η, it follows that the first of these terms is zero and so we see that

$$\int_a^b \left\langle \left(F_Y - \frac{d}{dt} F_Z \right), \eta \right\rangle dt = 0.$$

Finally, since this holds for *all* η satisfying the boundary conditions, an application of Lemma 11.2.1 completes the proof. □

Thus, in general here, we have a system of second-order ordinary differential equations (ODEs) with variable but smooth coefficients. As such, it also has a unique local solution (Appendix B) through each point and depends smoothly on the initial conditions.

The calculation of $J'(0)$, used to derive the Euler–Lagrange equations, is called the *first variation* of J. We now calculate $J''(0)$ called the *second variation* of J.

Proposition 11.2.3. *At* (t, Y, Z), *the second derivative,* $J''(0)(\eta)$, *is given by*

$$\int_a^b \sum_{i,j=1}^n \left[F_{z_i, z_j} \eta_i' \eta_j' + 2 F_{y_i, z_j} \eta_i \eta_j' + F_{y_i, y_j} \eta_i \eta_j \right] dt$$

for all η *satisfying* $\eta(a) = 0 = \eta(b)$.

Proof. Consider the second-order Taylor expansion of F,

$$F(t, Y + s\eta, Y' + s\eta')$$

$$= F(t, Y, Y') + s(F_{y_1}\eta_1 + \cdots + F_{y_n}\eta_n + F_{z_1}\eta_1' + \cdots + F_{z_n}\eta_n')$$

$$+ \frac{s^2}{2!} \left(\sum_{i,j=1}^n \frac{\partial^2 F}{\partial z_i \partial z_j} \eta_i' \eta_j' + 2\Sigma_{i,j=1}^n \frac{\partial^2 F}{\partial y_i \partial z_j} \eta_i \eta_j' \right.$$

$$\left. + \sum_{i,j=1}^n \frac{\partial^2 F}{\partial y_i \partial y_j} \eta_i \eta_j \right) + O(s^3).$$

We integrate this expression dt over $[a, b]$ and then differentiate under the integral sign with respect to s as above. This gives us

$$\frac{d}{ds}J(Y^*) = \int_a^b \frac{d}{ds}F(t, Y^*, Z^*)dt$$

$$= \int_a^b \frac{d}{ds}(F(t, Y, Z))dt + \int_a^b \frac{d}{ds}s(\langle F_Y, \eta \rangle + \langle F_Z, \eta' \rangle)dt$$

$$+ \int_a^b \frac{d}{ds}\frac{s^2}{2!}\sum_{i,j=1}^n [F_{z_i, z_j}\eta_i'\eta_j' + 2F_{y_i, z_j}\eta_i\eta_j' + F_{y_i, y_j}\eta_i\eta_j]dt$$

$$+ \int_a^b \frac{d}{ds}O(s^3)dt.$$

Thus,

$$\frac{d}{ds}J(Y^*) = \int_a^b (\langle F_Y, \eta \rangle + \langle F_Z, \eta' \rangle)dt + \int_a^b s\sum_{i,j=1}^n [F_{z_i, z_j}\eta_i'\eta_j'$$

$$+ 2F_{y_i, z_j}\eta_i\eta_j' + F_{y_i, y_j}\eta_i\eta_j]dt + \int_a^b O(s^2)dt.$$

Differentiating with respect to s a second time and then taking the limit as $s \to 0$ gives the result. $\sqcup$

From Proposition 11.2.3, we get a *necessary* condition for a minimum or a maximum.

Corollary 11.2.4. *If we have a minimum, or a maximum at Y, respectively, then the second variation is positive semidefinite (respectively negative semidefinite). Conversely, if the second variation is positive semi-definite (respectively negative semidefinite) and Y is a critical point, then it is a minimum (respectively maximum).*

Proof. We see from Proposition 11.2.3 that if a minimum occurs at Y, then for any η satisfying the boundary conditions and any small real s, since the Euler–Lagrange equations are satisfied by Y, the first derivative is zero. Hence,

$$J(Y + s\eta) = J(Y) + \frac{s^2}{2!}J''(0)(\eta) + O(s^3).$$

Therefore, Y is a minimum if and only if $\frac{s^2}{2!}J''(0)(\eta) + O(s^3) \geq 0$. Since $\frac{s^2}{2!} \geq 0$, this occurs if and only if $J''(0)(\eta) + O(s) \geq 0$. Taking the limit as $s \to 0$ shows this occurs if and only if $J''(0)(\eta) \geq 0$. The case of a maximum is similar. $\square$

As a corollary, we have the following sufficiency condition.

Corollary 11.2.5. *Let J and $F(t, Y, Z)$ be as above with F is independent of Y and suppose the $n \times n$ symmetric matrix, $\frac{\partial^2 F}{\partial z_i \partial z_j}$, is positive semidefinite everywhere (respectively negative semidefinite everywhere). If Y is a critical point of J, then Y is a global minimum (respectively maximum) for J.*

Proof. Since F is independent of Y and so $F_Y = 0$, it follows that

$$\frac{\partial^2 F}{\partial Y \partial Z} = 0 = \frac{\partial^2 F}{\partial Y \partial Y}.$$

Hence, if $\frac{\partial^2 F}{\partial z_i \partial z_j}$ is positive semidefinite everywhere, so is the second variation. $\square$

Exercise 11.2.6. Involving the Legendre condition.[3] Prove that if the second variation is positive semidefinite, then $\sum_{i,j=1}^{n} F z_i, z_j \eta_i' \eta_j'$ is also positive semidefinite and similarly for negative semidefinite.

Suggestion: Let $v = (v_1, \ldots, v_n) \in \mathbb{R}^n$ be arbitrary and $t_0 \in (a, b)$. Choose a small neighborhood N_0 of t_0 inside (a, b). Let $\eta_v : [a, b] \to \mathbb{R}^n$ be a linear tent function supported on N_0 and $\eta_v(t_0) = \epsilon v$. Then each η_v is continuous and takes the proper boundary values and η_v' is 0 outside of N_0 and equals $\pm v$ depending on whether we are to the left or right of t_0. Then $J''(0)(\eta_v) = \frac{1}{2\epsilon} \int_{t_0-\epsilon}^{t_0+\epsilon} \sum_{i,j=1}^{n} F z_i, z_j v_i v_j dt + O(\epsilon) \geq 0$. Then let $\epsilon \to 0$. If one doesn't like these tent functions and wants smooth η instead, one can just approximate the tent function by a C^∞ function and since the integrals will approximate one another, we get the same conclusion with smooth η.

---

[3] Adrien Legendre (1752–1833) was a French mathematician who made numerous important contributions to mathematics, such as the Legendre polynomials, Legendre transformation and the method of least squares. His name is inscribed on the Eiffel Tower.

We shall illustrate the use of Theorem 11.2.2 and Corollary 11.2.5 with two examples: one from geometry and one from mechanics.

11.2.3   Geodesics on a Riemannian surface

A smooth manifold is a space which is locally diffeomorphic with an open ball in $\mathbb{R}^n$. Here, n is fixed and is called the dimension of the manifold. When $n = 2$, the space is called a surface and it will be surfaces with which we will be concerned here. Here is Riemann's idea in general: a Riemannian manifold M is a connected (so that any two points can be joined by some smooth curve) and complete (basically, M has no holes or missing points) manifold with the property that at each point there is a positive definite matrix of order n, the dimension of the manifold and these matrices vary smoothly from point to point in the sense that each of their $i, j = 1, \ldots, n$ coordinates is a smooth numerical function on the space M. The purpose of these positive definite symmetric matrices, or positive definite quadratic forms, is to measure infinitesimal distances, i.e., the length of a very small arc passing through a point. If $(g_{i,j}(p))$ is the matrix at a point p and $X(t)$ for $-\epsilon \leq t < \epsilon$ is a small piece of curve passing through p at, say $t = 0$, then for $\epsilon > 0$ but small, its length should be roughly the length of its tangent vector to the curve at p. So, if $(u_1, \ldots, u_n)$ coordinatizes a small neighborhood of p in which the curve lies (and which is called a chart at p), then by the chain rule,

$$X'(t) = \sum_{i=1}^{n} X_{u_i}(u_i(t), \ldots, u_n(t)) \frac{du_i}{dt}.$$

So, $\|X'(t)\|$ should be

$$\sqrt{\sum_{i=1,j}^{n} \langle X_{u_i}, X_{u_j} \rangle \frac{du_i}{dt} \frac{du_j}{dt}}.$$

This is what $(g_{i,j}(p))$ are for. We define $\langle X_{u_i}, X_{u_j} \rangle$ to be the given $g_{i,j}(p)$. We then integrate these infinitesimal lengths to calculate the global length of an ordinary smooth arc joining two points on M. (Note that these positive definite symmetric matrices of Riemann are not given by anything having to do with M being embedded in some larger space). Thus, we say that these are intrinsic to M. The best ideas are sometimes easily stated!

In the case of a surface, the terms of the symmetric matrix are usually denoted E, F and G and so the metric given by

$$ds^2 = E(u,v)du^2 + 2F(u,v)dudv + G(u,v)dv^2.$$

We will now take the reader on a brief tour of the geometry of surfaces of constant curvature. To illustrate the definition of a Riemannian surface, we give two examples. One is the Poincare upper half plane, H^+, which consists of points (x, y) in the plane where $y > 0$. Here, the metric or positive definite symmetric matrix at (x, y) is a multiple of the identity, namely, $\frac{1}{y^2}I$. It is clearly positive definite and smoothly varying over H^+. This is an important example of a Riemannain surface. It is simply connected and has constant negative curvature. The form of this metric is said to be *conformal* to that of Euclidean space as it is a (variable) positive multiple of it. What this means is that the angles between intersecting curves are the same as the Euclidean angles, but the distances will not be and explains somehow why the curvature is a negative constant while that of Euclidean space is identically zero!

We will now find the geodesics in H^+. To do so, we first note what the hyperbolic metric is in relation to the Euclidean metric which we write as $ds_{Euc} = \sqrt{dx^2 + dy^2}$. Thus, $ds_{Hyp} = \frac{ds_{Euc}}{y}$ (which means the hyperbolic metric blows up as one approaches the boundary). Integrating, we see that the hyperbolic length s of a curve is given by the formula

$$s = \int \frac{\sqrt{1 + (y'^2)}}{y}dx.$$

Now, let P and Q be points in H^+ connected by some geodesic curve. If P and Q lie above one another, then the geodesic cannot be given by $y = f(x)$, nor can s be given by this formula, rather the curve must be a function of y. Here, it is is easy to see what the geodesic distance is since $P = (a, b)$ and $Q = (a, c)$ with, say $c > b$. For then

$$s = \int_b^c \frac{\sqrt{1 + (y'^2)}}{y}dy \geq \int_b^c \frac{1}{y}dy = \log(c) - \log(b),$$

and the inequality can be an equality only if y' is identically zero. That is, y is constant, so $c = b$! This proves that if P and Q lie on a vertical line, then that line is indeed a geodesic and its hyperbolic length is $\log(c) - \log(b)$.

Now, suppose P and Q do not lie above one another. Then the formula above is the operative one and then $F(x, y, y') = \frac{\sqrt{1+(y'^2)}}{y}$. Since F is independent of x, we know $F - y'F_{y'} = c_0$ a constant. From this, it easily follows that $1 = cy\sqrt{1 + (y')^2}$. Solving for the derivative, we get $\frac{dy}{dx} = \frac{\sqrt{c^2 - y^2}}{y}$, where $c = \frac{1}{c_0}$. Separating the variables and integrating gives $-2x + 2\alpha = 2\sqrt{c^2 - y^2}$, where 2α is a constant of integration. Hence, there are constants α and β so that $(x - \alpha)^2 + y^2 = \beta^2$. Thus, the geodesic is a semicircular arc between them, whose center lies on the x-axis, or put another way, the circle intersects the boundary of H^+ orthogonally. We remark it is evident from Euclidean geometry that given P and Q not lying above one another there is such a unique semicircle. It is also evident from Euclidean geometry that if they do lie above one another, there can't be such a circle since the construction to locate its center on the x-axis will fail because in Euclidean space, parallel lines can never meet!!

In the upper half plane, we have shown that, just as in the Euclidean plane itself, any two points of the hyperbolic plane can be joined by a unique geodesic and of minimal length. The significance of knowing all the geodesics is that here given a point and a geodesic ("straight line") through it there are infinitely many geodesics which don't meet it no matter how far extended within H^+. Hence, here, as opposed to Euclidean space, there are *infinitely many* lines in H^+ parallel to a given line.

We remark that sometimes curves on a Riemannian manifold are called "geodesics" if they merely satisfy the Euler–Lagrange equations, *but may not have minimal length*. By contrast to the upper half plane, in the case of the sphere, we will see an example of a surface where there are more than one (actually infinitely many) minimal length geodesics joining certain points and, in the case of the cylinder, a surface where there are infinitely many "geodesics" of larger and larger lengths joining a pair of fixed points. Of course, this also occurs in the sphere when a geodesic goes around a great circle more than once.

The reader might also be interested in another useful way of viewing the hyperbolic plane, i.e., as the interior of the open unit disk D. Note that D is also diffeomorphic to $\mathbb{R}^2$ (Prove). The hyperbolic metric on D is given by $ds_{\text{Hyp}} = \frac{ds_{\text{Euc}}}{(1-r^2)^2}$, where the Euclidean coordinates on D are taken to be its polar coordinates, (r, θ). D also has a boundary and as r approaches it, the hyperbolic metric blows up as before. This way of viewing hyperbolic space has circular symmetry as well as a distinguished point, namely, the center and these features can be exploited as in the following exercise.

Exercise 11.2.7.

1. Find all geodesics in D. In particular, show that all geodesics are arcs of circles (or straight lines through the origin), which intersect the boundary orthogonally.
2. Show the geodesics passing through the origin O if and only if they are straight lines.
3. Show that the Cayley transform extends to the boundary of H^+ and takes it onto the boundary of D.
4. What can one say about the sum of the interior angles of any geodesic triangle?

The way to go from the upper half plane H to D is by the *Cayley transform*, $c(z) = \frac{z-i}{z+i}$. Then, note the following.

Exercise 11.2.8.

1. c is well defined.
2. c is a diffeomorphism of H with D.
3. c is actually an isometry of H with D.
4. c extends to the boundary of H and takes it onto the boundary of D.

The Cayley[4] transform is a special case of the *Riemann mapping theorem* (see [9]). Here is an interesting consequence of the fact that H and D are isometric. c takes geodesics of H to those of D and

---

[4]Arthur Cayley (1821–1895) FRS was a British mathematician and eventually a professor at Cambridge who worked widely in mathematics, including group theory, linear algebra, complex analysis, non-Euclidean geometry and higher-dimensional geometry. He published over 900 papers during his lifetime.

therefore takes geodesic triangles of H to those of D. Now, just looking at a geodesic triangle of D and drawing the Euclidean triangle
determined by its vertices, we see that since the sum of the interior
angles of the Euclidean triangle is π, the concavity of these circles
(or lines) forces the sum of the interior angles of the geodesic triangle
to be $< \pi$.

A second example, or rather a class of examples, to analyze are
surfaces of revolution. These are isometric to submanifolds of $\mathbb{R}^3$ and,
as we know, can be constructed as follows. (Here, we are making use
of the ambient space.) For $a \leq x \leq b$, let $y = g(x)$ be a smooth
curve in the x, y plane, where $g(x) \geq 0$ and consider the surface in
$\mathbb{R}^3$ obtained by rotating the curve about the x-axis. The resulting
surface of revolution is defined by the equation $g(x)^2 - y^2 - z^2 = 0$.
For example, by taking the top half of a circle for g, we would get a
sphere, or by taking a horizontal line, we would get a cylinder. (If we
took an inclined line, we would get a cone, but this is not a smooth
manifold since it has a singularity at the vertex. We could also take
a whole circle lying above the x-axis and then get the surface of a
bagel. As an exercise, we ask the reader to work out the specifics of
the metric out in this case). Now, we introduce surface parameters
(which make it locally diffeomorphic to a ball in $\mathbb{R}^2$ at every point).
Let $a \leq u \leq b$ and $0 \leq v \leq 2\pi$. Then the surface of revolution consists
of those points in $\mathbb{R}^3$ of the form $X(u,v) = (u, g(u) \cos v, g(u) \sin v)$.
A simple calculation shows that $X_u = (1, g'(u) \cos v, g'(u) \sin v)$ and
$X_v = (0, g(u) - \sin v, g(u) \cos v)$. Hence, $\langle X_u, X_u \rangle = 1 + g'(u)^2 = E$,
$\langle X_u, X_v \rangle = F = 0$ (which is not usually the case, but when it is, it
is very helpful) and $\langle X_v, X_v \rangle = g(u)^2 = G$. So, here, $g_{i,j}$ is diagonal
with entries $1 + g'(u)^2$ and $g(u)^2$ and gives a smoothly varying positive
definite symmetric matrix at each point of the surface. In the case of
the sphere, there are singularities at the two poles because the metric
is not positive definite there, but nowhere else. As we shall see, this
is just a technicality which will not cause us trouble.

Now, let P and Q be distinct points on the surface of the unit
sphere, S^2. For example, NY and Paris and imagine being an airline
pilot flying from P to Q. To save fuel, one would want to find the
shortest route. Consider P, Q and the center O of the earth. These
form a triangle and generate a plane whose intersection with S^2 is
an arc of a great circle. Actually, there are two arcs: a long one and
a short one. We will show that the short one is the shortest path

on S^2 from P to Q. The only thing that can go wrong here is if P and Q happen to be diametrically opposite points on S^2. Then P, Q and O do not form a triangle because they are colinear and so don't determine a plane. Indeed, there are infinitely many planes passing through this line and each gives two geodesic paths from P to Q. But in any case, there will always be a great circle path, but this time not a unique one! We now will show that a great circle path is the shortest and at the same time study geodesics on surfaces of revolution more generally.

The Euler–Lagrange equation for the functional $ds^2 = E(u,v)du^2 + 2F(u,v)dudv + G(u,v)dv^2$ on a Riemannian surface is a system of two equations in terms of the parameter t of the curve. However, if in the parameter domain, we consider a curve of the form $v = v(u)$ or $u = u(v)$, then these give curves on the surface whose infinitesimal length is $ds = \sqrt{Eu'^2 + 2Fu' + G}$ or $ds = \sqrt{E + 2Fv' + Gv'^2}$, respectively, and whose integral dv, respectively du, gives the length function as at the end of Chapter 6. $s = \int \sqrt{Eu'^2 + 2Fu' + G}dv$, respectively $s = \int \sqrt{E + 2Fv' + Gv'^2}du$. Therefore, in order to minimize the length (with fixed end points) the Euler–Lagrange equation for one or the other of these functionals must be satisfied. In this way, we have reduced the problem from having to deal with a system of second-order equations to that of a single equation. However, a moments reflection tells us that this equation will not be easy to solve without some simplifying assumptions. For example, in general, it undoubtedly involves the independent variable with respect to which we are integrating, so the method which has worked so well for us by getting what is actually a first-order equation will not work in general. But suppose, for example, that $g_{i,j}(u,v)$ depend on u alone. Then we consider the second of these functionals and corresponding Euler–Lagrange equations. The situation is now much simpler. That is, $\frac{d}{du}\frac{F+Gv'}{\sqrt{E+2Fv'+Gv'^2}} = 0$. Hence, for some constant c,

$$F + Gv' = c\left(\sqrt{E + 2Fv' + Gv'^2}\right).$$

If in addition the diagonal term F is identically zero (which means that up on the surface, the curves $u = $ a constant and $v = $ a constant are orthogonal), the situation becomes even simpler and we have

$Gv' = c\left(\sqrt{Gv'^2 + E}\right)$. Solving for $v' = \frac{E}{G^2 - c^2 G}$ and integrating in the usual manner gives

$$v(u) = c \int \frac{\sqrt{E}\,du}{\sqrt{G^2 - c^2 G}}.$$

Of course, by symmetry, if $g_{i,j}(u,v)$ depended on v alone and $F = 0$, then considering the first of these Euler–Lagrange equations where $u = u(v)$, we would get

$$u(v) = c \int \frac{\sqrt{G}\,dv}{\sqrt{E^2 - c^2 E}}.$$

Such things have been extensively investigated in the 19th century and are called Clairaut surfaces.

We now apply these notions to study the geodesics on a surface of revolution. Here, $ds^2 = (1 + g'(u)^2)du^2 + g(u)^2 dv^2$ and, as we observed, $E = 1 + g'(u)^2$, $F = 0$ and $G = g(u)^2$. Here, the matrix coefficients are independent of v, so, as above, using the second equation, if we have a curve $v = v(u)$ in the parameter plane, the Euler–Lagrange equation becomes

$$v = c \int \frac{\sqrt{1 + g'(u)^2}}{g(u)\sqrt{|g(u)|^2 - c^2}}\,du.$$

We may now suppose that the generating curve has been parameterized by arc length, which we again call u. This is a normalization condition and every smooth curve can always be reparameterized in this way. Then $1 + g'(u)^2 = 1$ for all u. As an exercise, the reader should show that this is equivalent to parameterization by arc length. The reader should also prove that every smooth curve can always be reparameterized by arc length. Then

$$\frac{dv}{du} = \frac{c}{g(u)\sqrt{|g(u)|^2 - c^2}}.$$

There are two possibilities. If $c = 0$, then $\frac{dv}{du} = 0$ and so v is constant. These are exactly the plane sections through the x-axis of the surface and are always geodesics for any surface of revolution. The others require investigation.

In the case of a cylinder, $g(u) = c_0 > 0$. Hence, $v = c \int \frac{1}{c_0 \sqrt{|c_0|^2 - c^2}} du$. Thus, v is a linear function of u. The only other possibility is u is itself constant. These give circles transverse to the axis of the cylinder, and in this case, these are also clearly geodesics. When v is constant, we have straight lines parallel to the axis of the cylinder. When we have other linear functions, these give all helices. Thus, these are also "geodesics". But given any two points on the cylinder which have different u and v coordinates, we can find *infinitely many* helices joining them and exactly one of these will have minimal length.

The surface of a sphere of radius $r > 0$, can be viewed as a surface of revolution by taking $g(u) = \sqrt{r^2 - u^2}$. We conclude from the above observations on surfaces of revolution in general that great circles through the north pole N (on the axis of rotation) are geodesics. But then since the group of rotations takes any point on the sphere to N, it follows that any great circle through any point is a geodesic. Moreover, that is all. For if we had any geodesic, it would satisfy $\frac{dv}{du} = \frac{c}{g(u)\sqrt{|g(u)|^2 - c^2}}$, where g is as above. Assume that the geodesic passes through a point P. By applying the rotation group, always assume that it passed through N. Hence, its derivative $\frac{dv}{du}(0) = 0$ since it would have to lie in the tangent plane at N. But $\frac{c}{g(u)\sqrt{|g(u)|^2 - c^2}} \neq 0$ unless $c = 0$. Hence, $c = 0$ and the geodesic is a great circle. Thus, on the sphere, the geodesics are exactly the great circles. (In the exercises, we outline another way to deal with this using something analogous to Lagrange multipliers.)

The significance of knowing the geodesics here is that we now know any two points can be joined by a geodesic and if the surface distance is less than πr, the geodesic is unique. This means that a pilot flying from New York to Paris knows exactly which path to take if it is to be the shortest. However, if the distance is πr, then there are infinitely many geodesics (a continuum) joining these points. This is because then they are antipodal points and the great circle or plane is not uniquely determined, since the center of the earth lies on the line joining these points and contributes nothing to determining the plane. The spheres of various radii are the compact *simply connected* surfaces of constant (positive) curvature. Any two geodesics ("straight lines") must meet. Hence, here (where the curvature is positive) there are no parallel lines!

11.2.4   The principle of least action

The principle of least action or Hamilton's[5] principle is as follows: Let us consider a system of N particles in a space of dimension d. Let $n = dN$ and we consider the configuration space of the system to be $\mathbb{R}^n$. That is, these N particles are each running around in a space of dimension d. Let $q_1, \ldots, q_n$ be the position coordinates of points in the system and $\dot{q}_1, \ldots, \dot{q}_n$ be their respective instantaneous velocity (or momentum) coordinates. Denote by $U = U(t, q_1, \ldots, q_n, \dot{q}_1, \ldots, \dot{q}_n)$ the potential energy and by $K = K(t, q_1, \ldots, q_n, \dot{q}_1, \ldots, \dot{q}_n)$ the kinetic energy of the system, respectively. Then $E = K + U$ is the total energy and $L = K - U$ the Lagrangian. (Here, $-U$ is the work done on the system.) We consider paths $Q(t) = (q_1(t), \ldots, q_n(t))$ between $t = a$ and $t = b$, which minimize the action, $\int_a^b L\,dt$. (This is the principle of least action.) For such paths, we have $L_{q_i} - \frac{d}{dt} L_{\dot{y}_i} = 0$ for all $i = 1, \ldots, n$. The most important (and tractable!) case is when $U = U(q_1, \ldots, q_n)$ is a function of position alone (a conservative potential) and K is a quadratic form in the velocity coordinates, $K = \frac{1}{2} \sum_{i,k=1}^n a_{i,k} \dot{q}_i \dot{q}_k$, where $a_{i,k}$ is a real symmetric matrix. (Here, the Lagrangian is independent of t.) In this case, since each $U_{\dot{q}_i} = 0$, $L_{\dot{q}_i} = K_{\dot{q}_i}$ and therefore by Euler's theorem on homogeneous functions, $\sum_{i=1}^n \dot{q}_i L_{\dot{q}_i} = \sum_{i=1}^n \dot{q}_i K_{\dot{q}_i} = 2K$. Hence,

$$\sum_{i=1}^n \dot{q}_i L_{\dot{q}_i} - L = 2K - (K - U) = K + U = E.$$

Let us now calculate $\frac{dE}{dt} = \frac{d}{dt}\left(\sum_{i=1}^n \dot{q}_i L_{\dot{q}_i} - L\right)$. But this is

$$\sum_{i=1}^n \dot{q}_i \frac{d}{dt} L_{\dot{q}_i} + L_{\dot{q}_i} \ddot{q}_i - \frac{dL}{dt}.$$

By the Euler–Lagrange equations, this is

$$\sum_{i=1}^n \dot{q}_i L_{q_i} + L_{\dot{q}_i} \ddot{q}_i - \frac{dL}{dt}.$$

---

[5]Sir William Hamilton (1805–1865) was an Irish mathematician, astronomer, and physicist. He invented quaternions and was the Andrews Professor of Astronomy at Trinity College Dublin and also Royal Astronomer of Ireland.

Since $L = K - U$, $\frac{dL}{dt} = \sum_{i=1}^{n} \dot{q}_i L_{q_i} + L_{\dot{q}_i} \ddot{q}_i$ and so $\frac{dE}{dt} = 0$! This proves the principle of conservation of energy in a much more general form.

Corollary 11.2.9. *When the potential energy of the system is conservative and the kinetic energy is quadratic, then the path of least action conserves the total energy of the system throughout its trajectory.*

Let us now assume one thing further, namely, the symmetric matrix is actually diagonal and the diagonal entries correspond to the respective masses $m_1, \ldots, m_N$ of the N particles in the following manner. The n diagonal entries of the matrix are $m_1, \ldots, m_1$, $m_2, \ldots, m_2$, and finally, $m_N, \ldots, m_N$, each with multiplicity $\frac{n}{N} = d$. Then $K = \frac{1}{2} \sum_{i=1}^{N} m_i \|v_i\|^2$, where each $\|v_i\|^2$ is the sum of the squares of the d components of $\dot{q}_i^2$. On the other hand, by definition, $U_{q_i} = F_{q_i}$, the force in the q_i-direction and this is $m_i \ddot{q}_i$. It follows that for each $i = 1, \ldots, N$, i.e., for each particle, we have $F_i = m_i a_i$. This is Newton's second law of motion. We now consider the $2n \times 2n$ matrix which represents the second variation for this problem. The block in the lower corner is the symmetric matrix $\frac{\partial^2 L}{\partial \dot{q}_i \partial \dot{q}_j}$. In our case, since U is conservative, this is just A. On the other hand, $L_{q_i} = U_{q_i}$ therefore $L_{q_i, \dot{q}_j} = 0$. Hence, the second variation is block diagonal and consists of the blocks U_{q_i, q_j} and A. If these are positive semidefinite for all values on the trajectory, then by Theorem 11.2.4, the path which solves the system of Euler–Lagrange equations will indeed give the least action. Now, in the case where A represents the masses, it is indeed positive definite. In the case of a potential which depends linearly on the coordinates, $\frac{\partial^2 U}{\partial q_i \partial q_j} = 0$, and so the second variation here is positive semidefinite.

11.3 Vector-Valued Curves in Dimension n with up to n Constraints

In this section, we present a general theorem on extrema with constraints which is an analog of the theorem on Lagrange multipliers in calculus of several variables. In fact, its proof uses that result!

Applications when $k = 1$ of course include the problem of the hanging cable as well as other problems involving constraints, such as the isoperimetric problem.

Theorem 11.3.1. *For a curve Y in $\mathbb{R}^n$, let $J(Y) = \int_a^b F(x, Y, Y')dx$ be a functional we want to extremize, subject to constraints*

$$K_i(Y) = \int_a^b G_i(x, Y, Y')dx = c_i,$$

where $i = 1, \ldots, k$. If $k \leq n$, then there exist λ_i, where $i = 1, \ldots, k$ such that a solution Y satisfies the Euler–Lagrange equation for

$$F^* = F - \sum_{i=1}^{k} \lambda_i G_i \ (F^* \text{ is called the Lagrangian}).$$

Proof. Here, the k Lagrange multipliers are to be determined. For $j = 1, 2$, let $h_j(x) = (h_{j,1}(x), \ldots, h_{j,n}(x))$ be arbitrary smooth functions on $[a, b]$ with values in $\mathbb{R}^n$ satisfying $h_j(a) = 0 = h_j(b)$ and $s_j = (s_{j,1}, \ldots, s_{j,n}) \in \mathbb{R}^n$. Holding h_1 and h_2 fixed, consider

$$\Psi(s_1, s_2) = J(F^*(x, Y + s_1 h_1 + s_2 h_2, Y' + s_1 h_1' + s_2 h_2')).$$

Thus, Ψ is a function of $2n$ real variables, $s_{1,1}, \ldots, s_{1,n}, s_{2,1}, \ldots, s_{2,n}$ and we want to calculate its partial derivatives, $\frac{\partial \Psi(s_1, s_2)}{\partial s_{j,l}}$ at $(s_1, s_2) = (0, 0)$. However, these variables $s_{j,l}$ are not all independent. There must be relations among them because, throughout this process (i.e., for all s_1 and s_2), we shall insist that the $K_i(Y + s_1 h_1 + s_2 h_2)$ be held at the constant value c_i and this must hold for all $i = 1, \ldots, k$. Thus, the number of independent variables is actually only $2n - k$. This is where the k Lagrange multipliers come in. They boost the number of independent variables back up to $2n$.

The proof now follows the general pattern of that of Theorem 11.2.2. To calculate these partial derivatives we apply the first-order Taylor expansion to F^* at the point (x, Y, Y') and then integrate the result. For each j, l we get

$$\frac{\partial \Psi(s_1, s_2)}{\partial s_{j,l}} = \int_a^b \left(\frac{\partial F^*}{\partial Y} \frac{\partial Y}{\partial s_{j,l}} + \frac{\partial F^*}{\partial Y'} \frac{\partial Y'}{\partial s_{j,l}} \right) dx,$$

$$F^*(x, Y + s_1 h_1, Y' + s_2 h_2)$$

$$= F^*(x, Y, Y') + F^*_{y_1} h_{1,1} s_{1,1} + \cdots + F^*_{y_n} h_{1,n} s_{1,n}$$

$$+ F^*_{z_1} h'_{2,1} s_{2,1} + \cdots + F^*_{z_n} h'_{2,1} s_{2,n} + O(s^2).$$

$$= \int_a^b \left(\frac{\partial F^*}{\partial Y_j} h_{j,l} + \frac{\partial F^*}{\partial Y'_j} h'_{j,l} \right) dx.$$

We now calculate $\int_a^b \frac{\partial F^*}{\partial Y'_j} h'_{j,l} dx$ using integration by parts, since

$$\frac{d}{dx}(F^*_{Y'_j} h_{j,l}) = \frac{d}{dx}\left(F^*_{Y'_j}\right) h_{j,l} + F^*_{Y'_j} h'_{j,l}.$$

Upon integrating, we get

$$(F^*_{Y'} h_{j,l})|_a^b = 0 = \int_a^b \frac{d}{dx}(F^*_{Y'}) h_{j,l} + \int_a^b F^*_{Y'} h'_{j,l}.$$

Hence, $\int_a^b F^*_{Y'} h'_{j,l} = - \int_a^b \frac{d}{dx}(F^*_{Y'}) h_{j,l}$, and therefore,

$$\frac{\partial \Psi(s_1, s_2)}{\partial s_{j,l}} = \int_a^b \left(F^*_Y - \frac{d}{dx}(F^*_{Y'}) h_{j,l} \right) dx.$$

Thus, if Y is an extreme value of the functional associated with F^*, then since $0 = \frac{\partial \Psi(s_1, s_2)}{\partial s_{j,l}}$ evaluated at $s_{j,l} = 0$, we see that $\int_a^b (F^*_Y - \frac{d}{dx}(F^*_{Y'}) h_{j,l}) dx = 0$, and this holds for all $h_{j,l}$. By the fundamental lemma and the fact that this holds for all j, l, this means that $F^*_Y - \frac{d}{dx}(F^*_{Y'}) = 0$.   □

11.4   Several Independent Variables

In this section, we deal with variational problems in several independent real variables. As we shall see, we still have necessary conditions for an extrema of our functionals, namely, the Euler–Lagrange equation(s). However, now, these will be PDEs rather than ODEs. Since this is only an introduction to the subject, just as we did for Riemannian manifolds, when we get down to the details, we will restrict our attention to the case of functions of two independent variables. Much of what happens in more variables is informed by the current hypothesis, although naturally some aspects are genuinely more complicated. We shall also only consider numerical functions.

11.4.1   Necessary conditions for an extrema

Let F be a smooth function of $2n + 1$ variables

$$F = F(x_1, \ldots, x_n, z, z_{x_1}, \ldots, z_{x_n}),$$

and we consider the functional $J(z) = \int \ldots \int_\Omega F dx_1 \ldots dx_n$, where

$$z = z(x_1, \ldots, x_n)$$

is a smooth function of the n independent real variables $x = (x_1, \ldots, x_n)$ which vary over a bounded domain Ω in $\mathbb{R}^n$. We assume that the domain has a smooth (or perhaps a piecewise smooth boundary, $\partial\Omega$) (whatever it takes to make the integral converge and Stokes theorem to be true see Appendix C). We want to extremize J subject to the condition that z takes pre-assigned smooth values on $\partial\Omega$. In order to do this, we proceed more or less just as in previous cases, namely, we consider a function $h(x_1, \ldots, x_n)$ which is defined on the domain and its boundary is smooth on the domain and identically zero on the boundary. We then calculate the derivative of J at z in the direction h.

Theorem 11.4.1. *For h as above,*

$$J'(z)(h) = \int \ldots \int_\Omega \left(F_z - \sum_{i=1}^n \frac{\partial}{\partial x_i} F_{z_{x_i}} \right) h(x_1, \ldots, x_n) dx_1 \ldots dx_n.$$

Lemma 11.4.2. *Let f be a continuous function $f : \Omega \to \mathbb{R}$ and suppose $\int_\Omega f(x)h(x)dx = 0$ (where $x = (x_1, \ldots, x_n)$) for all smooth functions $h : \Omega \to \mathbb{R}$ satisfying $h \equiv 0$ on $\partial\Omega$. Then $f \equiv 0$ on Ω.*

Proof.   The proof of this lemma is similar to that of Lemma 11.2.1. Let $x_0 \in \Omega$ be a point where $f(x_0) \neq 0$. We may assume $f(x_0) > 0$. The proof when $f(x_0) < 0$ is entirely similar. Choose a neighborhood U of x_0 in Ω where on U the function is $\geq \frac{1}{2}f(x_0)$. Choose a smooth non-negative function h on Ω which vanishes outside $\overline{U}$ and takes the value 1 at x_0. Then evidently h vanishes on $\partial(\Omega)$, and also, $\int_\Omega f(x)h(x)dx = \int_{\overline{U}} f(x)h(x)dx > 0$. This is a contradiction. □

A consequence of this result together with Lemma 11.4.2 gives the analog of the Euler–Lagrange equation, but this time it is a PDE.

Corollary 11.4.3. *If the function z is an extreme value for J, then*

$$F_z - \frac{\partial}{\partial x_i} F_{z x_i} = 0.$$

The corollary follows because if z is an extreme value of J, and so by the theorem,

$$\int \cdots \int_\Omega \left(F_z - \sum_{i=1}^n \frac{\partial}{\partial x_i} F_{z x_i} \right) h \, dx_1 \ldots dx_n = 0$$

for every smooth function h vanishing on the boundary. An application of Lemma 11.4.2 completes the proof.

For the proof of the theorem, we shall need the following lemma. Here, we refer to Urysohn's Lemma in Appendix C to get test functions.

Lemma 11.4.4. *If $f_1, \ldots f_n$ are smooth functions on Ω which vanish on the boundary, then*

$$\int \cdots \int \sum_{i=1}^n \frac{\partial f_i}{\partial x_i} dx_1 \ldots dx_n = 0.$$

Proof. This follows immediately from the general version of Stokes' theorem for domains in $\mathbb{R}^n$ in Appendix C because the integral on the right side of the equation is definitely zero. □

Proof of the theorem is as follows.

Proof.   Here, we abbreviate $(z_{x_1}, \ldots, z_{x_n})$ by z_x. For h as above,

$$J(z+h) - J(z) = \int \cdots \int_\Omega (F(x, z+h, z_x) - F(x, z, z_x)) dx_1 \ldots dx_n.$$

Applying the Taylor theorem to the integrand yields

$$F(x, z+h, z_x) - F(x, z, z_x) = F_z h + \sum_{i=1}^n F_{z_{x_i}} h_{x_i} + O(||h||^2).$$

Hence, $\frac{J(z+h) - J(z)}{||h||} \to 0$, as $||h|| \to 0$. Thus,

$$J'(z)(h) = \int \cdots \int_\Omega \left(F_z h + \sum_{i=1}^n F_{z_{x_i}} h_{x_i} \right) dx_1 \ldots dx_n.$$

Now,

$$\frac{\partial}{\partial x_i} \left(F_{z_{x_i}} h \right) = F_{z_{x_i}} h_{x_i} + h \frac{\partial}{\partial x_i} F_{z_{x_i}}.$$

Hence,

$$\sum_{i=1}^n \frac{\partial}{\partial x_i} (F_{z_{x_i}} h) = \sum_{i=1}^n F_{z_{x_i}} h_{x_i} + h \sum_{i=1}^n \frac{\partial}{\partial x_i} F_{z_{x_i}}.$$

But then,

$$J'(z)(h) = \int \cdots \int_\Omega \left(F_z - \sum_{i=1}^n \frac{\partial}{\partial x_i} F_{z_{x_i}} \right) h \, dx_1 \ldots dx_n$$

$$+ \int \cdots \int_\Omega \sum_{i=1}^n \frac{\partial F_{z_{x_i}} h}{\partial x_i} dx_1 \ldots dx_n.$$

Now, since $h \equiv 0$ on the boundary, so is $F_{z_{x_i}} h$ for each $i = 1, \ldots, n$. Hence, by Lemma 11.4.4,

$$\int \cdots \int \sum_{i=1}^n \frac{\partial F_{z_{x_i}} h}{\partial x_i} = 0.$$

Thus,

$$J(z)(h) = \int \cdots \int_\Omega \left(F_z - \sum_{i=1}^{n} \frac{\partial}{\partial x_i} F_{z_{x_i}} \right) h \, dx_1 \ldots dx_n$$

for every smooth function h vanishing on $\partial\Omega$. □

We now apply this last result to the problem of minimizing the area of a surface subject to fixed boundary conditions, namely, all such surfaces have the same boundary curve. This is called the *Plateau problem*.[6] It is named after Plateau because of his experiments with soap bubbles which, due to surface tension, gives a physical representation of the solutions to this problem for various boundary conditions. It was finally solved in 1930 independently by Tibor Rado (1895–1965) and Jesse Douglas (1897–1965). However, their solutions could have singular points. It was Robert Osserman (1926–2011) who finally in 1970 proved that, in their arguments, no singularities occurred. Note that the problem of geodesics joining two fixed points is a lower-dimensional analog of this problem.

Here, we deal with smoothness conditions. We shall make other more significant assumptions in the sequel.

Let Ω be a bounded region in the plane with smooth boundary, $\partial(\Omega)$, and $f(u,v)$ be a smooth real-valued function defined on Ω. We take the associated embedded surface $X(u,v) = (u, v, f(u,v))$. Then $X_u = (1, 0, f_u)$ and $X_v = (0, 1, f_v)$, so $E = 1 + f_u^2$, $F = f_u f_v$ and $G = 1 + f_v^2$. Since $\det g_{i,j} = 1 + f_u^2 + f_v^2$, we see that the infinitesimal area of this surface is $dA = \sqrt{1 + f_u^2 + f_v^2}$. Hence, the area of the surface is given by

$$A(\mathrm{Graph}(f)) = \int_\Omega \sqrt{1 + f_u^2 + f_v^2} \, du \, dv.$$

Thus, taking $F(u, v, f, f_u, f_v) = \sqrt{1 + f_u^2 + f_v^2}$, the variational problem of minimizing the area of all such surfaces passing through

---

[6] Joseph Plateau (1801–1883) was a Belgian physicist and mathematician who studied capillary action and surface tension. He invented an early version of a stroboscopic device.

a given space curve lying directly above $\partial(\Omega)$, we get the Euler–Lagrange equation,

$$\frac{\partial F}{\partial f} - \frac{\partial}{\partial u}\frac{\partial F}{\partial f_u}\frac{\partial}{\partial v}\frac{\partial F}{\partial f_v} = 0.$$

Since here F is independent of f, $\frac{\partial F}{\partial f_u} = \frac{f_u}{\sqrt{1+f_u^2+f_v^2}}$, and $\frac{\partial F}{\partial f_v} = \frac{f_v}{\sqrt{1+f_u^2+f_v^2}}$, we see that

$$\frac{\partial \frac{f_u}{\sqrt{1+f_u^2+f_v^2}}}{\partial u} + \frac{\partial \frac{f_v}{\sqrt{1+f_u^2+f_v^2}}}{\partial v} = 0,$$

from which it follows that

$$f_{uu}(1 + f_v^2) + f_{vv}(1 + f_u^2) - 2f_u f_v f_{uv} = 0. \tag{11.1}$$

In the late 18th century, Lagrange had observed that the mean curvature $H(u, v)$ at a point on a surface within $\mathbb{R}^3$ is given by exactly the left-hand side of this PDE. This means that given a smooth simple closed curve in $\mathbb{R}^3$, if we seek a smooth surface of minimal area in $\mathbb{R}^3$ with this curve as its boundary, the surface must satisfy this PDE. That is, minimal surfaces (ones where the area is minimal) are exactly those with mean curvature identically 0. This second-order PDE is an elliptic equation. As such, it has a unique global solution if Ω is simply connected. However, in general, the solution cannot be written explicitly since it depends both on Ω and the given curve lying above the boundary. Even when these are given, it is not at all easy to pin it down. Doing so would mean solving the Plateau problem. For some further details concerning elliptic PDEs, we refer the reader to Chapter 10.

Now, let X and Y be vectors in $\mathbb{R}^3$ and define the differential form $\omega(X, Y) = \det(X, Y, N)$, where N is the unit normal to Graph(f), namely,

$$N = \frac{(-f_u, -f_v, 1)}{\sqrt{1 + f_u^2 + f_v^2}}.$$

The cohomology based on these forms is called De Rham cohomology. Fortunately, what we are doing here only requires the following lemma and the fact that in our case, the cohomology is trivial and not extensive knowledge of this subject.

Lemma 11.4.5. ω *is called a De Rham 2 cocycle, i.e.,* $d(\omega) = 0.$

For a point (x, y, z) in $\mathbb{R}^3$, we have the following

Proof.

$$\omega\left(\frac{\partial}{\partial x}, \frac{\partial}{\partial y}\right) = \frac{1}{1 + \sqrt{1 + f_x^2 + f_y^2}} \det S,$$

where

$$S = \begin{pmatrix} 1 & 0 & -f_x \\ 0 & 1 & -f_y \\ 0 & 0 & 1 \end{pmatrix}$$

so that

$$\omega\left(\frac{\partial}{\partial x}, \frac{\partial}{\partial y}\right) = \frac{1}{1 + \sqrt{1 + f_x^2 + f_y^2}}.$$

Similarly,

$$\omega\left(\frac{\partial}{\partial x}, \frac{\partial}{\partial z}\right) = \frac{-f_x}{1 + \sqrt{1 + f_x^2 + f_z^2}}$$

and

$$\omega\left(\frac{\partial}{\partial y}, \frac{\partial}{\partial z}\right) = \frac{f_y}{1 + \sqrt{1 + f_y^2 + f_z^2}}.$$

Thus,

$$\omega = \frac{dx \wedge dy - f_x dy \wedge dz - f_y dz \wedge dx}{\sqrt{1 + f_x^2 + f_y^2}}.$$

Now,

$$d(\omega) = \frac{\partial}{\partial x}\left(\frac{-f_x}{\sqrt{1 + f_u^2 + f_v^2}}\right) + \frac{\partial}{\partial y}\left(\frac{-f_y}{\sqrt{1 + f_x^2 + f_y^2}}\right).$$

A *careful* calculation of the right side of this last equation shows $d(\omega)$ is a fraction whose numerator is the right side of (11.1) and whose denominator is $(1 + f_x^2 + f_y^2)^{3/2}$. Since f satisfies the minimal surface equation (11.1), this means that $d(\omega) = 0$. □

We will now show that $A(\text{Graph}(f))$ is minimizing among all surfaces Σ in the vertical cylinder, $\Omega \times \mathbb{R} \subseteq \mathbb{R}^3$ and the same boundary as $\text{Graph}(f)$.

Theorem 11.4.6. *If $f : \Omega \to \mathbb{R}$ satisfies* (11.1) *and $\Sigma \subseteq \Omega \times \mathbb{R}$ is any surface with $\partial(\Sigma) = \partial(\text{Graph}(f))$, then $A(\Sigma) \geq A(\text{Graph}(f))$*

Proof.   Let X and Y be orthonormal vectors at any point of $\mathbb{R}^3$. Then $|\omega(X, Y)| \leq 1$ with equality if and only if X, Y are tangent to the $\text{Graph}(f)$ at $(u, v, f(u, v))$. Since ω is a closed form, it is exact. Therefore, Σ and $\text{Graph}(f)$ are cohomologous. By Stokes' theorem,

$$\int_\Sigma \omega = \int_{\text{Graph}(f)} \omega.$$

Hence,

$$A(\text{Graph}(f)) = \int_{\text{Graph}(f)} \omega = \int_\Sigma \omega \leq A(\Sigma).$$

□

We note without proof the further development that when Ω is convex, this last result can be sharpened, namely, if f satisfies 11.1, then we get an *absolute* minimizing result. That is, $A(\text{Graph}(f)) \leq A(\Sigma)$ for all surfaces Σ with $\partial(\Sigma) = \partial(\text{Graph}(f))$.

We conclude Chapter 11 with some exercises and final remarks.

Exercise 11.4.7.

1. Consider the variational problem of geometric optics

$$J(y) = \int \frac{\sqrt{1 + y'^2}}{v(x, y)} dx,$$

where $v(x, y)$ is the velocity of light at the point (x, y) of the medium. Let $p(x, y) = \frac{1}{v(x, y)}$. Show that the Euler–Lagrange equation for

$$J(y) = \int p(x, y) \sqrt{1 + y'^2} dx$$

is a first-order equation.

2. Find all geodesics of a paraboloid of revolution, $z = x^2 + y^2$. Show that each geodesic on the paraboloid which lies on a plane section through the axis of rotation intersects itself infinitely often.

3. Show that in the case of the sphere and the hyperbolic plane, each geodesic is defined for all values of the parameter.

4. Consider the interior of the unit disk with the hyperbolic metric $ds^2 = \frac{ds^2_{euc}}{(1-(x^2+y^2))^2}$. Find all its geodesics. In particular, show geodesics always intersect the boundary orthogonally. Show the ones passing through 0 are straight lines. Show the sum of the angles of a geodesic triangle is strictly less than π. Show there are infinitely many lines parallel to a given line through a given point. This Riemannian surface is actually isometric to H^+.

5. Consider the variational problem where the functional

$$J(y) = \int_a^b p^2(x) y'^2 \, dx.$$

Show that if y is a critical point, then $y' = \frac{c}{p^2}$. Then show that the minimum value of J is $\frac{b-a}{\int_a^b p^{-2} dx}$.

6. Consider the variational problem where the functional J is obtained by integrating F and $F = F(x, y, y', \ldots y^n)$. Here, $n \geq 2$, and we are talking about the higher derivatives of y. Show the corresponding Euler–Lagrange equation is

$$F_y - \frac{d}{dx}(F_{\dot{y}}) + \frac{d^2}{dx^2}(F_{\ddot{y}}) + \cdots + (-1)^n \frac{d^n}{dx^n}(F_{y^n}).$$

Hint: Apply Taylor expansion of F.

7. Consider two functionals $J(y) = \int_a^b F(x, y, z) dx$ and

$$K(y) = \int_a^b G(x, y, z) dx.$$

Show the y which extremize J subject to the constraint $K = c$ also extremize K subject to the constraint $J = c$.

8. Formulate the variational problem of maximizing the volume of a figure closed in a surface of fixed area. What isoperimetric inequality does one get?

9. Suppose in a variational problem with a constraint the Lagrange multiplier turns out to be zero. What conclusion can you draw? The reader should formulate the corresponding result when $J(y) = \int F(x, y, y', y'')$ and there is a constraint. Then prove your statement.

10. Consider the problem of extremizing $\int_a^b (y\dot{})^2 dx$, where $y(a)$ and $y(b)$ take given fixed values subject to the constraint $\int_a^b y^2 dx = c$. Show that such a function y must be a solution to $\frac{d^2}{dx^2} y - \lambda y = 0$. Thus, here, the Lagrange multiplier is an eigenvalue of the operator $\frac{d^2}{dx^2}$.

11. Let M be a Riemannian surface with metric $ds^2 = E du^2 + 2F dudv + G dv^2$. Thus, if $X(t) = X(u(t), v(t))$, where $a \le t \le b$ is a curve in M, its length $L(X)$ is given as above by $L(X) = \int_a^b \sqrt{E u'^2 + 2F u'v' + G v'^2} dt$. We also define its energy by $E(X) = \frac{1}{2} \int_a^b (E u'^2 + 2F u'v' + G v'^2) dt$. Prove that $L(X) \le \sqrt{2(b-a)} \sqrt{E(X)}$, with equality if and only if the curve is parametrized by arc length. Thus, geodesics are exactly the energy minimizers. Note the relationship with Liapunov's theorem.

12. Let h be a smooth function on $[a, b]$ with $h(a) = 0$. Show

$$\int_a^b h^2(x) dx \le \frac{(b-a)^2}{2} \int_a^b (h'(x))^2 dx.$$

Hint: Since $h(a) = 0$, by the fundamental theorem of calculus, for all x $h(x) = \int_a^x h'(t) dt$. Squaring, integrating and applying the Schwarz inequality yields $\int_a^b h^2(x) dx \le \int_a^b (h'(t))^2 dt \int_a^b (x - a) dx$. Then note that $\frac{d}{dx} \frac{(x-a)^2}{2} = x - a$. Use this to show that $\int_a^b (\phi(x) h^2(x) + \psi(x) h'(x)^2) dx$ is a continuous functional of the functions ϕ and ψ in the sup norm.

13. Here, we outline an approach to dealing with a constraint on the manifold rather than on functionals. We will want to solve problems where we restrict the locus to a submanifold of $\mathbb{R}^n$ which is the zero set of a smooth function $\Phi : \mathbb{R}^n \to \mathbb{R}$ or defined on an open subset of $\mathbb{R}^n$. Let $x = (x_1, \ldots, x_n) \in \mathbb{R}^n$ and consider the submanifold S of zeros of Φ. For example, the sphere of radius

$r > 0$ is the zero set of $\Phi(x) = x_1^2 + \cdots + x_n^2 - r^2$. As a result, we have the following theorem which we state without proof.

Theorem 11.4.8 (General Constraint). *For a curve Y in $\mathbb{R}^n$, let*

$$J(Y) = \int_a^b F(t, Y, Y')dt$$

be a functional we want to extremize on S. If Y does this, then there is a smooth function $\mu(t)$ defined on $[a, b]$ such that Y satisfies the Euler–Lagrange equation for

$$F^*(t, Y, Z) = F(t, Y, Z) + \mu(t)\Phi(Y).$$

Our purpose is to indicate how this result gives another approach to the problem of geodesics of the 2 sphere. Let $F^* = F + \mu(t)(x^2 + y^2 + z^2 - r^2)$, where $\mu(t)$ is an unknown smooth function and $F = \sqrt{x^2 + y^2 + z^2}$. Since we are doing this in $\mathbb{R}^3$, we get a system of three equations

$$\mu(t)\frac{\partial \Phi}{\partial x} - \frac{d}{dt}\frac{\dot{x}}{F^*} = 0,$$

$$\mu(t)\frac{\partial \Phi}{\partial y} - \frac{d}{dt}\frac{\dot{y}}{F^*} = 0,$$

$$\mu(t)\frac{\partial \Phi}{\partial z} - \frac{d}{dt}\frac{\dot{z}}{F^*} = 0.$$

With this F^* and Φ, eliminating $\mu(t)$, we get

$$\frac{F\ddot{x} - \dot{x}\dot{F}}{2xF^2} = \frac{F\ddot{y} - \dot{y}\dot{F}}{2yF^2} = \frac{F\ddot{z} - \dot{z}\dot{F}}{2zF^2}.$$

Put another way,

$$\frac{y\ddot{x} - x\ddot{y}}{y\dot{x} - x\dot{y}} = \frac{\dot{F}}{F} = \frac{z\ddot{y} - y\ddot{z}}{z\dot{y} - y\dot{z}}.$$

Equating the last two tells us that

$$\frac{\frac{d}{dt}(y\dot{x} - x\dot{y})}{y\dot{x} - x\dot{y}} = \frac{\frac{d}{dt}(z\dot{y} - y\dot{z})}{z\dot{y} - y\dot{z}}.$$

This integrates to give $\log(y\dot{x} - x\dot{y}) = \log(z\dot{y} - y\dot{z}) + \log c$. Therefore, $y\dot{x} - x\dot{y} = c(z\dot{y} - y\dot{z})$. But then, dividing by y and solving for $\frac{\dot{y}}{y}$

yields $\frac{\dot{x}+c\dot{z}}{x+cz} = \frac{\dot{y}}{y}$. Integrating again gives $\log(x+cz) = \log(y) + \log c_1$ and so $x + cz - c_1 y = 0$. This is a plane through the origin, i.e., through the center of the sphere and therefore intersects the sphere in a great circle. Which planes, i.e., which great circles, are unaccounted for? The only ones unaccounted for are planes of the form $ay + bz = 0$, the ones passing through the x-axis. Hence, we don't get the geodesics joining $(r, 0, 0)$ with $(-r, 0, 0)$. But since, as discussed above, the rotation group carries such a pair of points to different pair of antipodal points and preserves spherical distance and geodesics, we see that the geodesics on a sphere are precisely the great circles.

We remark that this method's success depends on a certain symmetry between F and Φ. For example, the reader should verify that there is trouble if we simply replace the sphere by an ellipsoid. This is just the transitivity of the rotation group in a somewhat different guise.

Chapter 12

The Gauss Bonnet Theorem for Surfaces in $\mathbb{R}^3$

In calculus of 1 variable, one studies curvature of plane curves. Here, we do the analogous thing for surfaces S in $\mathbb{R}^3$, but as the reader might imagine, this will take much more doing. For example, instead of the tangent line at a point p of the curve, here we have the tangent plane $T_p(S)$ to the surface at p. Nevertheless, we will end up with powerful and interesting results. As we shall see, the assumption (following Gauss!) of the surface being embedded in $\mathbb{R}^3$ will be extremely helpful. Also, as we shall see at the end, the theorem of turning tangents (Umlaufsatz) of a simple closed plane curve will lie at the heart of the matter when we prove the global Gauss–Bonnet theorem.

12.1 Local Theory; Surfaces in $\mathbb{R}^3$

Let $X : U \times V \to \mathbb{R}^3$ be a smooth function, where U and V are intervals in $\mathbb{R}$ in the sense of [10] 1 variable. $U \times V$ is called the parameter space and its image $X(U \times V) = S$ the corresponding surface. Here, our treatment will be strictly local so that actually there is no distinction between intrinsic geometry and the geometry of an embedded surface. We denote the cross-product in $\mathbb{R}^3$ by $\times$. $X_u(p)$ and $X_v(p)$ denote the (vector) partial derivatives of $X(u, v)$ at the point p, which evidently lie in $T_p(S)$. We also write the inner product of vectors Y and Z in $\mathbb{R}^3$ by (Y, Z) and the corresponding norm by $\|Y\| = \sqrt{(Y, Y)}$.

The resulting proposition is an easy consequence of the inverse function theorem together with properties of the cross-product and is left to the reader an an exercise. The following equivalent conditions will constitute our further assumption at all points $p \in S$, called non-singularity, which we make from now on.

Proposition 12.1.1. *The following conditions are equivalent*:

1. $X_u(p) \times X_v(p) \neq 0$.
2. $X_u(p)$ and $X_v(p)$ are linearly independent.
3. X is a local diffeomorphism at p.

Evidently, from condition 3, we see that a diffeomorphic smooth change of parameter preserves non-singularity.

This is probably a good time to give some examples of embedded surfaces S (or X).

Example 1. Let f be a smooth numerical function of two real variables (u, v). Then $X(u, v) = (u, v, f(u, v))$ is such a surface. Since $X_u = (1, 0, f_u)$ and $X_v = (0, 1, f_v)$, we see that $X_u \times X_v = (\ f_u, f_v, 1)$, which is non-zero. (Indeed, since we are only going to be dealing with local properties, the theory of ODE tell us that this is actually the most general case!) A trivial example is when $f(u, v) = a + bu + cv$, where a, b, c are contents, and u and v range over all of $\mathbb{R}$. Here, of course, S is a plane. Of particular interest is the hyperboloid, $X(u, v) = (u, v, v^2 - u^2)$, where u and v also range over all of $\mathbb{R}$.

Example 2. Surfaces of revolution. Let $f(u)$ and $g(u)$ be any smooth functions of u defined on $[a, b]$ and $X(u, v) = (g(u)\cos(v), g(u)\sin(v), f(u))$, where $a \leq u \leq b$ and $0 \leq v \leq 2\pi$. Thus, $g(u)$ and $f(u)$ are the parametric equations of a curve in the x, z plane which is then rotated about the z-axis. Some specific examples of surfaces of revolution are the sphere, the right circular cone (excluding the vertex), the right circular cylinder, the torus and the tractrix.

Example 3. Consider the unit sphere with center $(0, 0, 1)$ which is tangent to the u, v plane at the origin. From the north pole

$n = (0, 0, 2)$, draw the straight line to $(u, v, 0)$. This line intersects the sphere at n and at

$$X(u, v) = \left(\frac{4u}{u^2 + v^2 + 4}, \frac{4v}{u^2 + v^2 + 4}, \frac{2(u^2 + v^2)}{u^2 + v^2 + 4} \right).$$

Giving a diffeomorphism from $S^2 \setminus \{n\}$ with $\mathbb{R}^2$ called steriographic projection.

Now, let $t \mapsto (u(t), v(t))$ be a smooth curve in parameter space, where $X(u(0), v(0)) = p$. Then $X((u(t), v(t))$ is a curve on S passing through p. By condition 3 above, locally all curves on S through p arise in this way. By the chain rule,

$$X'(t) = X_u u'(t) + X_v v'(t),$$

and so

$$\|X'(t)\|^2 = \|X_u\|^2 \, u'(t)^2 + 2(X_u, X_v) u'(t) v'(t) + \|X_v\|^2 \, v'(t)^2.$$

Letting $(g_{i,j})$ denote (X_{u_i}, X_{u_j}) for $i, j = 1, 2$, we get a 2×2 real matrix which is symmetric since the inner product is. Since the inner product is continuous and X is smooth, this matrix varies continuously as one moves from point to point. It is called the *first fundamental form*. Thus, we have a way of measuring infinitesimal distances in $T_p(S)$. Sometimes, it is convenient to use the notation,

$$\begin{bmatrix} E & F \\ F & G \end{bmatrix}.$$

Proposition 12.1.2. *The symmetric matrix, $(g_{i,j})$, is positive definite.*

Proof.

$$\|X_u \times X_v\|^2 = \|X_u\|^2 \, \|X_v\|^2 - (X_u, X_v)^2 = \det(g_{i,j}),$$

and since $X_u \times X_v \neq 0$, $\det(g_{i,j})$ is positive. Moreover, $g_{1,1} = \|X_u\|^2$ which is also positive because $X_u \neq 0$. As $(g_{i,j})$ is a 2×2 matrix, it is positive definite. $\square$

In this way, we get a formula for the length of a local curve through p on S, namely,

$$s = \int_0^b \sqrt{\sum_{i,j=1}^2 g_{i,j}(t)u_i(t)u_j(t)}dt.$$

Moreover, because infinitesimally the area of the parallelogram spanned by X_u and X_v is $\|X_u \times X_v\| = \sqrt{\det g_{i,j}}$, we also get a formula for area,

$$A = \iint_S \sqrt{\det(g_{i,j})(u,v)}dudv.$$

Here is how the first fundamental form is affected by a non-singular change of parameter J. We leave its proof which follows from the chain rule to the reader.

Proposition 12.1.3. *Let* $(u,v) \mapsto (\bar{u}, \bar{v})$ *be a smooth change of parameter and J be its derivative. Then*

$$(g_{i,j})(u,v) = J(g_{i,j})(\bar{u}, \bar{v})J^t.$$

The changed first fundamental form remains positive definite symmetric.

Proof. JAJ^t is symmetric since $(JAJ^t)^t = JA^tJ^t = JAJ^t$. Its determinant is $(\det J)^2 \det A$ which is positive since $\det A$ is. Diagonalizing A and letting λ_1 and λ_2 be its eigenvalues, one calculates

$$\mathrm{tr}(JAJ^t) = (a^2 + c^2)\lambda_1 + (b^2 + d^2)\lambda_2,$$

where

$$J = \begin{bmatrix} a & b \\ c & d \end{bmatrix}.$$

Since $\lambda_i > 0$, $\mathrm{tr}\, JAJ^t \geq 0$. If $\mathrm{tr}\, JAJ^t = 0$, then clearly $J = 0$, a contradiction. Therefore, $\mathrm{tr}\, JAJ^t > 0$ and because $\det JAJ^t$ is also positive and we are dealing with 2×2 matrices, it follows easily that JAJ^t is positive definite. $\square$

Example 12.1.4. Here, we calculate the first fundamental form in some examples.

Let $X(u, v) = (g(u)\cos(v), g(u)\sin(v), f(u))$ be a surface of revolution. Then

$$X_u(u, v) = (g'(u)\cos(v), g'(u)\sin(v), f'(u))$$

and

$$X_v(u, v) = (-g(u)\sin(v), g(u)\cos(v), 0).$$

Then one sees easily that $(X_u, X_v) = 0$, $\|X_u\|^2 = g'^2 + f'^2$, and $\|X_v\|^2 = g^2$. Thus, the first fundamental form is

$$\begin{bmatrix} g'^2 + f'^2 & 0 \\ 0 & g^2 \end{bmatrix}.$$

If $X(u, v) = (u, v, f(u, v))$, then the first fundamental form is

$$\begin{bmatrix} \sqrt{1 + f_u^2} & f_u f_v \\ f_u f_v & \sqrt{1 + f_v^2} \end{bmatrix}.$$

In particular, if $f(u, v) = a + bu + cv$ is linear, then

$$\begin{bmatrix} \sqrt{1 + b^2} & bc \\ bc & \sqrt{1 + c^2} \end{bmatrix},$$

which is constant (independent of p).

When $X(u, v) = (u, v, v^2 - u^2)$, we get

$$\begin{bmatrix} \sqrt{1 + 4v^2} & -4uv \\ -4uv & \sqrt{1 + 4u^2} \end{bmatrix}.$$

Exercise 12.1.5. Consider the surface of revolution generated by revolving the curve $v = u^4$ about the v-axis. Find its first fundamental form. Also, find its eigenvalues.

As opposed to Gaussian geometry, Riemannian geometry starts with a pre-assigned positive definite matrix $(g_{i,j})(p)$ for each $p \in S$ which varies smoothly with p, whereas what we have done above, following Gauss, is to construct first fundamental form $(g_{i,j}(p))$ from the embedding, X. This brings us to the following important distinction.

Definition 12.1.6. An intrinsic property is one which depends only on the first fundamental form and not on the embedding, X.

Exercise 12.1.7. Use the chain rule to show that if $(u, v) \mapsto (\bar{u}, \bar{v})$ is a smooth change of parameter and J is its derivative, then

$$\|X_{\bar{u}} \times X_{\bar{v}}\| = \frac{1}{|\det J|} \|X_u \times X_v\|.$$

12.2   Geodesics

Let $X'(s) = X_u u'(s) + X_v v'(s)$ be a curve on S parametrized by arc length $0 \le s \le l$. A geodesic on S is a curve on S joining $X(0)$ and $X(b)$ of shortest length,

$$L(X) = \int_0^b \sqrt{\sum_{i,j=1}^{2} g_{i,j} X_i'(s) X_j'(s)} ds.$$

As we know from Chapter 11, such curves must satisfy the Euler equations,

$$\Phi_u(s) - \frac{d}{ds} \Phi_{u'}(s) = 0,$$

$$\Phi_v(s) - \frac{d}{ds} \Phi_{v'}(s) = 0,$$

where $\Phi(s) = \sqrt{\sum_{i,j=1}^{2} g_{i,j} X_i'(s) X_j'(s)}$.

Generally, one considers a geodesic to be a curve parametrized by arc length, satisfying these Euler equations rather than a curve of shortest length (which is among them!). However, the square root in this integrand is inconvenient. For this reason, we consider the *Energy* functional,

$$E(X) = \int_0^l \frac{1}{2} \|X'(s)\|^2 ds,$$

instead.

That is, we will show that when seeking geodesics, we can replace Φ above by the Φ associated with the energy, namely,

$$\Phi = \frac{1}{2} \sum_{i,j=1}^{2} g_{i,j} X_i'(s) X_j'(s),$$

and use the Euler equations of this new Φ instead. We emphasize that arc length is always the parameter.

We first observe that if $\Psi(u, v, u', v')$ satisfies the Euler equations and $\Phi(u, v, u', v') = \frac{1}{2}\Psi^2 i(u, v, u', v')$, then so does Φ.

Proof. For

$$\frac{\partial \Phi}{\partial u} = \Psi \frac{\partial \Psi}{\partial u},$$

and since arc length is the parameter,

$$\Psi(s) = \|X'(s)\| = 1.$$

Therefore, $\frac{\partial \Phi}{\partial u} = \frac{\partial \Psi}{\partial u}$. Similarly,

$$\frac{\partial \Phi}{\partial u'} = \Psi \frac{\partial \Psi}{\partial u'}.$$

Hence,

$$\Phi_u(s) - \frac{d}{ds}\Phi_{u'}(s) = \Psi_u(s) - \frac{d}{ds}\Psi_{u'}(s).$$

Similarly,

$$\Phi_v(s) - \frac{d}{ds}\Phi_{v'}(s) = \Psi_v(s) - \frac{d}{ds}\Psi_{v'}(s). \qquad \square$$

Definition 12.2.1. Given an embedded surface $X : U \times V \to S \subseteq \mathbb{R}^3$, let

$$\nu(p) = \frac{X_u(p) \times X_v(p)}{\|X_u(p) \times X_v(p)\|}.$$

Since $\|X_u(p) \times X_v(p)\| \neq 0$, for each $p \in S$, we get a normal vector to the surface for each p, i.e., $\nu(p)$ is perpendicular to $T_p(S)$.

We now connect the Euler equations for the energy to the geometry of S by observing, $\Phi_u(s) - \frac{d}{ds}\Phi_{u'}(s) = 0$ if and only if $(X''(s), X_u(s)) = 0$ and $\Phi_v(s) - \frac{d}{ds}\Phi_{v'}(s) = 0$ if and only if $(X''(s), X_v(s)) = 0$.

Proof. Since $\Phi = \frac{1}{2}(X', X')$, $\Phi_u = (X'_u, X')$, while

$$\frac{d}{ds}\Phi_{u'}(s) = \frac{d}{ds}(X'_u(s), X'(s)) = (X''(s), X'_u(s)) + (X'(s), X'_u(s))$$

and similarly for v, v'. $\qquad \square$

As a result, we get the following.

Corollary 12.2.2. *$X(s)$ is a geodesic if and only if $(X''(s),$ $X_u(s)) = 0$ and $(X''(s), X_v(s)) = 0$. That is, $X''(s)$ is always perpendicular to $T_p(X(s))$ at every point on the curve. Hence, we write $X''(s) = k_\nu(s)\nu(s)$.*

Definition 12.2.3. k_ν is called the normal curvature (we will define the geodesic curvature shortly).

See Chapter 11 for the analysis of the geodesics on the various surfaces of revolution mentioned at the beginning of this section. We also found all geodesics in the Poincare upper half plane H^+ as well. It is worth remarking here that this calculation was based only on the hyperbolic metric and not on an embedding of H^+ in $\mathbb{R}^3$.

Exercise 12.2.4. Let $(r\cos(\theta), r\sin(\theta), v)$ coordinatize a cylinder of radius $r > 0$, where $0 \le \theta < 2\pi$ and $v \in \mathbb{R}$. Choose points $p_0 = (r, 0, 0)$ and $q_0 = (r\cos(\theta), r\sin(\theta), v)$, where $v > 0$ and $0 < \theta < \pi$. Show there are infinitely many geodesics joining p_0 and q_0, but only one of minimal length. Do the same for the cone.

Exercise 12.2.5. Use the appropriate integral inequality [10] to show

$$\int_0^l \sqrt{\sum_{i,j=1}^2 g_{i,j} X_i'(s) X_j'(s))^2} ds \le l \int_0^l \sum_{i,j=1}^2 g_{i,j} X_i'(s) X_j'(s) ds.$$

We conclude this section with the *compatibility relations* which we will need later. For this, see [8].

Proposition 12.2.6.

1. $(X_{u,u}, X_u) = \frac{1}{2} E_u.$
2. $(X_{u,u}, X_v) = F_u - \frac{1}{2} E_v.$
3. $(X_{u,v}, X_u) = \frac{1}{2} E_v.$
4. $(X_{u,v}, X_v) = \frac{1}{2} G_u.$
5. $(X_{v,v}, X_u) = F_v - \frac{1}{2} G_u.$
6. $(X_{v,v}, X_v) = \frac{1}{2} G_v.$

12.3   The Gauss Map

We formalize things by calling the map $p \mapsto \nu(p)$, the *Gauss map*. It takes $S \to S^2$. S is called *orientable* if this map is continuous, and in that case, it is actually smooth, and as we know $\nu(p)$ is never zero and is perpendicular to T_p at each point $p \in S$.

Proposition 12.3.1. *When arc length is the parameter, $X(s)$ is a geodesic if and only if $X''(s)$ is a a multiple of $\nu(s)$.*

Proof.   Since at each point the triple X_u, X_v, ν forms a basis of the tangent space extended to $\mathbb{R}^3$. If X is a geodesic, then $X''(s)$ is a multiple of $\nu(s)$ at each point on X and since $\|\nu(s)\| = 1$ everywhere, if $X(s)$ is a geodesic, since $\|\nu(s)\| = 1$, $\|X''(s)\| = k_\nu(s)$. Conversely, if

$$\|X''(s)\| = k_\nu(s) \text{ for all } s,$$

then $X''(s)$ must be a multiple of $\nu(s)$ and so X is a geodesic.   □

Now, let $X(s)$ be any curve on S parameterized by arc length (not necessarily a geodesic). Hence,

$$X''(s) = a(s)X_u(s) + b(s)X_v(s) + k_\nu(s)\nu(s).$$

We can now single out important intrinsic geometric invariants of X. First of all, when S is *orientable*, the Gauss map itself is intrinsic since ν is the *unique* outward pointing vector of unit length perpendicular to the tangent plane at each point. $\alpha(s) = a(s)X_u(s) + b(s)X_v(s)$ is called the curvature vector. Then

$$(X''(s), X_u(s)) = ag_{1,1} + bg_{1,2} \quad (X''(s), X_v(s)) = ag_{2,1} + bg_{2,2}.$$

Since $\det(g_{i,j})(s) \neq 0$ (in fact is greater than 0 as we are assuming S is orientable), we can solve for a, b:

$$a = \frac{g_{2,2}(X''(s), X_u(s)) - g_{1,2}(X''(s), X_v(s))}{\det(g_{i,j})(s)},$$

$$b = \frac{g_{1,1}(X''(s), X_v(s)) - g_{2,1}(X''(s), X_u(s))}{\det(g_{i,j})(s)}.$$

Proposition 12.3.2. *$a(s)$ and $b(s)$ are intrinsic and $X(s)$ is a geodesic if and only if $a(s) = 0 = b(s)$ for all s on the curve.*

This is because (X_u, X_v, ν) are linearly independent. Thus, X is a geodesic if and only if the curvature vector $\alpha(s) = 0$. If $\theta(s)$ is the angle between $X''(s)$ and $\alpha(s)$, then

$$\cos(\theta) = \frac{\|\alpha\|}{\|X''\|}.$$

Now, let $k_g(s) = \frac{\pm\|\alpha(s)\|}{X''(s)}$ taking $+$ or $-$ according to whether $0 \leq \theta \leq \frac{\pi}{2}$ or $0 \geq \theta \geq -\frac{\pi}{2}$. $k_g(s)$ is called the *geodesic* curvature.

Proposition 12.3.3.

1. $(\alpha(s), X'(s)) = 0$, *i.e.,* α *is a multiple of* X'.
2. $\|\nu(s) \times X'(s)\| = 1$.
3. $\alpha(s) = k_g(s)(\nu(s) \times X'(s))$.

Proof.

1. Since $X(s)$ is parametrized by arc length, $(X'(s), X'(s)) = 1$. Differentiating yields $(X'(s), X''(s)) = 0$. But $X'' = \alpha + k_{nu}\nu$ so that

$$(X'', X') = (\alpha, X') + (k_{nu}\nu, X') = (\alpha, X') = 0.$$

2. $\|\nu(s) \times X'(s)\|^2 = \|\nu(s)\|^2 \|X'(s)\|^2 - (\nu(s), X'(s))$. But since

$$(\nu(s), X'(s)) = 0, \|\nu(s)\| = 1 \text{ and } \|X'(s)\| = 1.$$

   The conclusion follows.
3. $(X'(s), \nu(s)) = 0$ since X' is in the tangent space. Hence, $(\alpha, \nu) = 0$. But also $(\alpha, X') = 0$. Therefore, $\alpha(s) = \lambda(s)(\nu(s) \times X'(s))$. Taking norms, we get $\|\alpha(s)\| = |\lambda(s)| \|\nu(s) \times X'(s)\| = |\lambda(s)| = \pm k_g(s)$.

$\square$

Corollary 12.3.4. $X''(s) = k_g(s)(\nu(s) \times X'(s)) + k_{\nu(s)}\nu(s)$. *In particular,* $X(s)$ *is a geodesic if and only if* $k_g(s)$ *is identically zero. In general,* $k_g(s) = (X''(s), \nu(s) \times X'(s))$ *and* $k_{nu(s)} = (X''(s), \nu(s))$.

Exercise 12.3.5. Show that if a curve $X(s)$ on S is projected orthogonally onto the tangent plane of any of its points, then the resultant plane curve has curvature equal to $k_g(s)$ for that s.

We conclude this section with Lioville's formula for k_g. Here, for simplicity, we shall assume that we have *orthogonal* coordinates, i.e., the first fundamental form has $F = 0$ everywhere. As usual, $\theta(s)$ is the angle between the given curve $X(s)$ and $X_u(s)$.

Proposition 12.3.6. *If $F = 0$ everywhere on an oriented surface S and $X(s)$ is a smooth curve in S, then*

$$k_g(s) = \frac{1}{2\sqrt{EG}}\left(G_u\frac{dv}{ds} - E_v\frac{du}{ds}\right) + \frac{d\theta(s)}{ds}.$$

Proof. Let $e_1 = \frac{X_u}{\sqrt{E}}$ and $e_2 = \frac{X_v}{\sqrt{G}}$. Then $e_1 \times e_2 = \nu$. Hence,

$$\left(\frac{de_1}{ds}, \nu \times e_1\right) = \left(\frac{de_1}{ds}, e_2\right) = ((e_1)_u, e_2)\frac{du}{ds}\,((e_1)_v, e_2)\frac{dv}{ds}.$$

Since $F = 0$, the equations of compatibility tell us $(X_{u,u}, X_v) = -\frac{1}{2}E_v$ and so $((e_1)_u, e_2) = (\frac{X_u}{\sqrt{E}}, \frac{X_v}{\sqrt{G}}) = -\frac{1}{2\sqrt{EG}}E_v$. Similarly, $((e_1)_v, e_2) = \frac{1}{2\sqrt{EG}}G_u$. $\qquad\square$

12.4   Gaussian Curvature: Second Fundamental Form

Here, we consider an oriented embedded surface $X : U \times V \to S$. Then at each fixed point $X(u_0, v_0)$, for a nearby point $X(u, v)$, we have

$$X(u, v) - X(u_0, v_0)$$
$$= (u - u_0)X_u((u_0, v_0)) + (v - v_0)X_v((u_0, v_0))$$
$$+ \frac{1}{2}[(u - u_0)^2 X_u, u((u_0, v_0)) + 2(u - u_0)(v - v_0)X_u, v((u_0, v_0))$$
$$+ (v - v_0)^2 X_v, v((u_0, v_0)) + \epsilon(u, v),$$

where $\frac{\epsilon(u,v)}{\|(u-u_0, v-v_0)\|} \to 0$ as $(u, v) \to (u_0, v_0)$.

Then the distance from $X(u, v)$ from the tangent plane at $p_0 = X(u_0, v_0)$ is the inner product, $(X(u, v) - X(u_0, v_0), \nu((u_0, v_0)))$, which by our calculation just above in the limit is

$$(u - u_0)(X_u((u_0, v_0), \nu) + (v - v_0)(X_v((u_0, v_0), \nu)$$
$$+ \frac{1}{2}[(u - u_0)^2(X_{u,u}((u_0, v_0), \nu) + 2(u - u_0)(v - v_0)$$
$$\times (X_{u,v}((u_0, v_0)\nu) + (v - v_0)^2(X_{v,v}((u_0, v_0), \nu).$$

The first two terms are zero since they lie in T_{p_0}. Letting $L = (X_{u,u}(u_0, v_0), \nu)$, $M = (X_{u,v}(u_0, v_0)\nu)$, $N = (X_{v,v}(u_0, v_0), \nu)$, then for $(u, v) \to (u_0, v_0)$, we see that $d(u, v)$ is approximately

$$\frac{1}{2}[(u - u_0)^2 L + 2(u - u_0)(v - v_0)M + (v - v_0)^2 N].$$

This quadratic form at each point $p \in S$ is called the second fundamental form. It is evidently non-intrinsic.

Exercise 12.4.1. Use the fact that nu is perpendicular to T_{p_0} at every point to show the following:

$$L(u, v) = -(\nu_u, X_u),$$
$$M(u, v) = (\nu_u, X_v) = (\nu_v, X_u),$$
$$N(u, v) = -(\nu_v, X_v).$$

Now, let $X(s) = X(u(s), v(s))$ be a curve on S parametrized by arc length. Then we know $k_\nu = (X''(s), \nu(s))$ and also

$$X''(s) = X_u u''(s) + X_v v''(s) + X_{u,u} u'(s)^2 + 2X_{u,v} u'v' + X_{v,v} v'(s)^2.$$

Hence,

$$k_\nu(s) = L(u(s), v(s))u'(s)^2 + 2M(u(s), v(s))u'(s)v'(s)$$
$$+ N(u(s), v(s))v'(s)^2.$$

When S is orientable, we know that k_ν is smooth. Since

$$Eu'(s)^2 + 2Fu'(s)v'(s) + Gv'(s)^2 = (X'(s), X'(s)) = 1$$

and since the first fundamental form is positive definite, we are studying the behavior of the second fundamental form on the ellipse:

$$Lu'(s)^2 + 2Fu'(s)v'(s) + Gv'(s)^2 = 1.$$

By compactness of this ellipse, the continuous function $k_{\nu(s)}$ has extreme values $k_{\max}$ and $k_{\min}$ (which could coincide). These are called the principal curvatures. To find them, we use Lagrange multipliers λ, the constraint being this ellipse.

$$(L - \lambda E)u + (M - \lambda F)v = 0,$$
$$(M - \lambda F)u + (N - \lambda G)v = 0.$$

Since the principal curvatures exist, this system of homogeneous linear equations is degenerate and so

$$\det \begin{bmatrix} L - \lambda E & M - \lambda F \\ M - \lambda F & N - \lambda G \end{bmatrix} = 0,$$

i.e.,

$$(EG - F^2)\lambda^2 + (LMF - EN - GL)\lambda + (LN - M^2) = 0,$$

where $k_{\max}$ and $k_{\min}$ are the roots. This is a quadratic equation since $EG - F^2 = \det(g_{i,j}) \neq 0$. We define $H = \frac{1}{2}(k_{\max} + k_{\min})$ which is called the *mean curvature* and $K = k_{\max}k_{\min}$ the *Gaussian curvature*. Note that

$$K = \frac{LN - M^2}{EG - F^2},$$

the quotient of the determinants of the second and first fundamental forms. For an embedded surface, this gives a convenient way of calculating K by looking at second derivatives. Since the equation above factors as

$$(\lambda - k_{\max})(\lambda - k_{\min}) = 0,$$

it follows that for all real λ, $\lambda^2 - 2H\lambda + K = 0$.

Evidently, the principal curvatures are equal if and only if $H^2 = K$. With the principal curvatures comes the *principal directions*. These are the vectors in the tangent plane T_{p_0} to the places on the ellipse where they are achieved. It is a remarkable fact proved by Gauss that in spite of the fact that the second fundamental form is not intrinsic, the Gaussian curvature is.

Exercise 12.4.2. Show that the principal curvatures are equal if and only if $\frac{L}{E} = \frac{M}{F} = \frac{N}{G}$. When this happens, one says p_0 is an umbilic.

Exercise 12.4.3. Show that $K_\nu = k_{\max} \cos^2(\theta) + k_{\min} \sin^2(\theta)$, where as usual θ is the angle between X' and X_u.

The curvature at p_0 gives us a good qualitative description of what S looks like locally near a point p_0.

Definition 12.4.4.

1. If $K(p_0) > 0$, then we say p_0 is an elliptic point. That is, $k_{\max}$ and $k_{\min}$ have the same (non-zero) sign. Here, locally, T_{p_0} lies on one side of the surface.
2. If $K(p_0) < 0$, then we say p_0 is a hyperbolic point. That is, $k_{\max}$ and $k_{\min}$ have opposite signs. Here, locally, the surface cuts the tangent plane and lies on both sides of it.
3. If $K(p_0) = 0$, then we say p_0 is a parabolic point. That is, $k_{\max}$ or $k_{\min}$ are zero. If both are zero, we say p_0 is a planar point. This case is locally somewhat like the elliptic case, but is degenerate.

Exercise 12.4.5. An example of 1 is any point on the sphere, and an example of 2 is the point $(0, 0)$ on the hyperbola $(u, v, v^2 - u^2)$. An example of 3 is a point on the cylinder or cone, and an example of a planar point is a point on a plane, but another is the point $(0, 0)$ on the surface of revolution generated by $v = u^4$. As an exercise, we ask the reader to verify these facts by calculating the Gaussian curvature.

Exercise 12.4.6. Consider the surface of revolution given by the torus. By calculation, show that points on the outside have positive Gaussian curvature, those on the inside have negative Gaussian curvature and those at the very top and bottom (these are points on the uppermost and lowermost circles) have zero curvature, the latter following from the first two statements by continuity of K.

12.5   The Exponential Map and Geodesic Polar Coordinates

Here, we are interested in special features of a well-chosen local coordinate system about a fixed point p_0, namely, one in which locally for each p in a neighborhood of p_0 satisfies $(X_u(p), X_u(p)) = 1$ and $(X_u(p), X_v(p)) = 0$. Since $\det(g_{i,j}) > 0$, this forces $(X_v(p), X_v(p)) > 0$. Calling that term $G = g^2(u, v)$, we see that length dominates Euclidean length here:

$$\int_0^l \sqrt{u'(t)^2 + g^2(u, v)v'(t)^2}\, dt > \int_0^l u'(t)\, dt = u(l) - u(0). \quad (12.1)$$

Such a curve is a geodesic, actually giving *shortest* length of all curves in S joining p_0 to other fixed, but nearby points p. We call such a neighborhood a *geodesic* coordinate system.

How can we produce such a local coordinate system? By the local existence theorem of ODE for each direction $v \in T_{p_0}$, there is a local geodesic $\gamma(t, v)$ through p_0 so that $\gamma'(0) = v$. Then reparametrize this curve by its arc length, s.

Lemma 12.5.1. *Suppose the geodesic* $\gamma(s, v)$ *is defined for all* $|s| < \epsilon$ *and* $c \neq 0 \in \mathbb{R}$. *Then the geodesic* $\gamma(s, cv)$ *is defined for all* $|s| < \frac{\epsilon}{c}$ *and* $\gamma(s, cv) = \gamma(cs, v)$.

Proof. Let α be defined on $(-\frac{\epsilon}{c}, \frac{\epsilon}{c})$ by $\alpha(s) = \gamma(cs)$. Then $\alpha(0) = \gamma(0)$ and $\alpha'(0) = c\gamma'(0) = c^2 v$. By local existence and uniqueness, α is a geodesic with initial conditions $\gamma(0)$ and $c\gamma'(0)$. Hence, $\alpha(s) = \gamma(s, cv) = \gamma(cs, v)$.   □

If $v \neq 0 \in T_{p_0}$ and has small enough norm, then $\gamma(\|v\|, \frac{v}{\|v\|}) = \gamma(1, v)$ is defined. We then define exp, the exponential map of differential geometry, by $\exp_{p_0}(v) = \gamma(1, v)$. We also take $\exp_{p_0}(0) = p_0$, our base point. In this way, for each point p_0, we have a well-defined function $\exp_{p_0}$ defined in a neighborhood V of 0 in T_{p_0}. By the fundamental theorem of ODE, $\exp_{p_0}$ is smooth since γ is. Indeed, $\exp_{p_0}$ is actually a diffeomorphism from V to its image in S. To see this, we compute its derivative. For each $v \in V$,

$$\frac{d}{ds} \exp_{p_0}(sv)|_{s=0} = \frac{d}{ds} \gamma(s, v)|_{s=0} = v.$$

This means that $d(\exp_{p_0})|_0 = I$, the identity. Since this linear operator is non-singular, the inverse function theorem [10] tells us $\exp_{p_0}$ is a diffeomorphism of some sub-neighborhood of V containing the zero vector which we again call V. On V, we have polar coordinates and since $\exp_{p_0}$ is a diffeomorphism, its inverse image gives polar coordinates on $U = \exp^{-1}(V)$. Here, U is a neighborhood of p_0 in S and the radial curves emanating from p_0 are geodesics. We call U a normal neighborhood of p_0 in S. Its coordinates called *geodesic polar coordinates* are (ρ, θ). The curves $\theta = $ a constant are geodesics.[1]

Proposition 12.5.2. *Let $E(\rho, \theta)$, $F(\rho, \theta)$ and $G(\rho, \theta)$ be the first fundamental form in these coordinates. Then $E = 1$, $F = 0$, $G > 0$. Moreover, $\lim_{\rho \to 0} G = 0$ and $\lim_{\rho \to 0} \sqrt{G}_\rho = 1$.*

In particular, since $F = 0$, this means that the curves ρ is constant intersect the curves θ is constant orthogonally. This is called *Gauss' Lemma*.

Proof. By definition of the exponential map, ρ is the arc length of the curve θ is constant. This means $E = 1$. We now apply the first of the compatibility relations 12.2.6. Since θ being constant is a geodesic, $0 = \frac{1}{2}E_\rho$. By the second of these relations, it follows that $F_\rho = 0$, and therefore, F depends only on θ. For each $p \in V$ (other than p_0), let $\rho = \phi(\theta)$ be its radial part, where $0 \leq \theta < 2\pi$. But there are radial geodesics $\gamma(s)$ passing through p (s being arc length). Then

$$F(\rho, \theta) = \left(\frac{d\phi}{d\theta}, \frac{d\gamma}{ds} \right).$$

Now, although $F(\rho, \theta)$ is not defined at $\rho = 0$ (namely, at p_0), if we fix θ, since $\phi(\theta)$ is constant, its derivative $\frac{d\phi}{d\theta} = 0$ and so the right

---

[1]However, note that to have a coordinate system, we require $\rho > 0$ and $0 < \theta < 2\pi$. Therefore, we have to remove the line $\theta = 0$. But this is harmless because removing this from U and its image from V still leaves a diffeomorphism.

side of the above equation is zero. Hence,

$$\lim_{\rho \to 0} F(\rho, \theta) = 0.$$

Since F is independent of ρ, this means $F = 0$. Therefore, $EG - F^2 = G$, and as we noted earlier, the first fundamental form is positive definite; therefore, $G = \det(g_{i,j}) > 0$.

Finally, by 12.1.7 under a smooth change of variable with derivative J, areas transform as follows: $\sqrt{EG - F^2} = \sqrt{E^*G^* - F^{*2}}|\det J|$. When $u^* = \rho\cos(\theta)$, $v^* = \rho\sin(\theta)$, then $|\det J| = \rho$. Hence, $\sqrt{G} = \rho\sqrt{E^*G^* - F^{*2}}$. But in the $*$ coordinates, we are in *Euclidean* space, so $E^* = 1 = G^*$ and $F^* = 0$. Thus, $\sqrt{G} = \rho$. This completes the proof. $\qquad\square$

From equation (12.1), we get the following.

Proposition 12.5.3. *Given* $p_0 \in S$, *there exists a neighborhood* U *about* p_0 *with the property that if* $p \in U$, *the unique geodesic joining* p_0 *to* p *has shortest length of any curve in* S *joining* p_0 *and* p.

Note that even though p_0 is technically excluded from the chart, K is certainly defined there.

Theorem 12.5.4. *Taking geodesic coordinates, we have* $K(\rho, \theta) = -\sqrt{G}_{\rho,\rho}/\sqrt{G}$ *(recall* $G > 0$ *everywhere). That is,* $\sqrt{G}$ *satisfies the ODE*

$$\sqrt{G(\rho, \theta)}_{\rho,\rho} + K\sqrt{G(\rho, \theta)} = 0.$$

In particular, if K is constant, then $\sqrt{G(\rho, \theta)}$ satisfies the second-order ODE in ρ, with constant coefficients.

We defer the proof of this theorem until the final section of this chapter. As the reader will see, we can draw some interesting consequences from it. First, we need to calculate the Taylor series of $g(\rho, \theta) = \sqrt{G}(\rho, \theta)$ as a function of ρ up to order 3.

Proposition 12.5.5. *Let* K_{p_0} *be the value of* K *at the origin. Then,*

$$g(\rho, \theta) = \rho - \frac{K_{p_0}}{3!}\rho^3 + \epsilon(\rho, \theta),$$

where $\lim_{\rho \to 0} \frac{\epsilon}{\rho^3} = 0$ *uniformly in* θ.

Proof.   Since $g((\rho,\theta)_{\rho,\rho} = -K(\rho,\theta)g(\rho,\theta)$, differentiating with respect to ρ, we get $\frac{\partial^3 g}{\partial\rho^3} = -Kg_\rho - gK_\rho$. But as $\rho \to 0$, $g \to 0$. Hence, $\lim_{\rho\to 0}\frac{\partial^3 g}{\partial\rho^3} = -Kg_\rho$. But since $\rho \to 0$, the limit of g_ρ is 1, and we see

$$\lim_{\rho\to 0}\frac{\partial^3 g}{\partial\rho^3} = -K(p_0). \tag{12.2}$$

Now, the Taylor expansion in ρ of the smooth function $g(\rho,\theta)$ is

$$g(\rho,\theta) = g(0,\theta) + \rho g_\rho(0,\theta) + \rho^2/2!g_{\rho,\rho}(0,\theta) + \rho^3/3!g_{\rho,\rho,\rho} + \epsilon(\rho,\theta).$$

Since the limiting value of each of the first and third terms is each zero (see [10]), this proves the result.   □

We can now calculate the arc length l of a geodesic circle of radius r. Since the points of discontinuity in geodesic polar coordinates is a set of measure zero,

$$l = \int_0^{2\pi} g(r,\theta)d(\theta).$$

Using Proposition 12.5.5, we get, as a limiting value,

$$l = 2\pi r - \frac{\pi}{3}r^3 K(p_0). \tag{12.3}$$

This equation gives an intrinsic way of understanding the effect of the curvature at a point in terms of arc length of nearby geodesic circles, the arc length in the tangent space T_{p0} of such a circle being $2\pi r$. Here, we get a correction term. If $K(p_0) > 0$, it makes l smaller. If $K(p_0) < 0$, it makes l bigger.

Solving for $K(p_0)$ yields $K(p_0) = \lim_{r\to 0}\frac{3}{\pi r^3}(2\pi r - l)$ which gives a way of calculating the curvature $K(p_0)$ from the arc length of small geodesic circles and proves that K is intrinsic, i.e., depends only on the first fundamental form (and its derivatives).

One can also calculate the area A of a geodesic circle of radius r. In general, this is $\int \sqrt{EG - F^2}dudv$ which in our case is

$$A = \int_0^r \int_0^{2\pi} g(\rho,\theta)d(\rho)d(\theta).$$

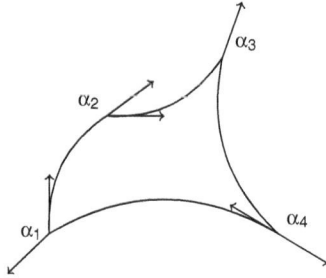

Figure 12.1. A simple piecewise smooth closed curve on a surface of variable curvature.

of positive, zero and negative curvature). The surfaces where $g \geq 2$ have to do with hyperbolic geometry and have negative curvature everywhere. The homology groups of S are as follows: $H^0(S, \mathbb{Z}) = \mathbb{Z}$, $H^1(S, \mathbb{Z}) = \mathbb{Z}^{2g}$ and $H^2(S, \mathbb{Z}) = \mathbb{Z}$. Thus, the Euler characteristic $\chi(S) = \sum_{i=0}^{2}(-1)^{i-1} \operatorname{rank}(H^i(S)) = 2 - 2g$. As we shall see, the Euler characteristic will play an important role in what follows.[2]

We now define the concept of a simple closed piecewise smooth curve.

Definition 12.6.1. By a piecewise smooth curve $X(s)$ on the surface S, we shall mean s is arc length and the total length is l. We assume that there are at most a finite number s_i where $i = 0, \ldots, n$ and that X is smooth on $[0, l]$ except at s_i, and at these points, the arcs meet at exterior angles α_i, where $0 < \alpha_i < 2\pi$. (Note that the corresponding *interior angles* $\theta_i = \pi - \alpha_i$.) At all other points of $X(s)$, we have a well-defined tangent $X'(s)$ and angle $\phi(s)$ (the angle between $X'(s)$ and $X_u'(s)$).

Definition 12.6.2. Such a curve $X(s)$, $s \in [0, l]$ is called a simple if it never crosses itself and is called closed if $X(0) = X(l)$.

---

[2]For all remarks concerning homology, the reader should consult any of the standard books on topology where it is also proved that such an S can be triangulated. Given a triangulated surface S, its Euler characteristic is defined as $v - e + f$, where v is the number of vertices, e the number of edges and f the number of faces in the triangulation. Obviously, we have in mind that the Euler characteristic should be independent of the triangulation. This is because it is equal to the alternating sum of the ranks of the homology groups as above. All this and more is true and is explained in many topology books.

Now, let $X(s) = X(u(s), v(s))$ be a simple closed piecewise smooth curve on the orientable surface $X : (U \times V) \to S$ parametrized by arc length. Since U and V can be taken to be finite open intervals. $U \times V$ is homeomorphic with an open disk in $\mathbb{R}^2$. Therefore, we can apply Green's theorem. Moreover, if $f(u, v)$ is a smooth function defined in the interior of $X(s)$, we can form the integral, $\iint f(u, v)\sqrt{EG - F^2}dudv = \iint f(u, v)dA$. We shall also assume that $X(s)$ is positively oriented, i.e., as one traverses it, the normal to the curve points upward. Then the intuitively obvious theorem on Turning Tangents or Umlaufsatz tells us that

$$\sum_{i=0}^{n} \phi(s_{i+1}) - \phi(s_i) + \sum_{i=0}^{n} \alpha_i = 2\pi.$$

We first turn to the *local* Gauss–Bonnet theorem.

Theorem 12.6.3. *Let* $U \times V$ *be a geodesic coordinate patch and* $X(s)$ *be as above. Then*

$$\iint K dA + \sum_{i=0}^{n} \int_{s_i}^{s_{i+1}} k_g(s)ds + \sum_{i=0}^{n} \alpha_i = 2\pi. \qquad (12.5)$$

We remark that *no other continuous function* $K^* : X(U \times V) \to \mathbb{R}$ can do this! For if K^* were such a function, then $\iint_S K dA = \iint_S K^* dA$ for any measurable subset $S \subseteq X(U \times V)$. So, for example, if there were a point p where $K(p) > K^*(p)$, then there would be a neighborhood

$$S \subseteq X(U \times V)$$

about p where this would remain the case, and therefore,

$$\iint_S K dA > \iint_S K^* dA,$$

a contradiction, and similarly if $K(p) < K^*(p)$.

As a corollary, we get the original form of the theorem, due to Gauss, which concerns the *interior angles*, θ_i. It gets us right back to the origins of Euclidean plane geometry!

Corollary 12.6.4. *Let $U \times V$ be as above and $X(s)$ consist of geodesics. Then*

$$\iint K dA + \sum_{i=0}^{n} \alpha_i = 2\pi.$$

In particular, if $X(s)$ is a geodesic triangle, then

$$\iint K dA = \theta_1 + \theta_2 + \theta_3 - \pi.$$

This latter quantity is called the excess. *Thus in any geodesic triangle T, the total curvature $\iint K dA =$ the excess. In particular, if $K > 0$ everywhere, the excess is positive; if $K < 0$ everywhere, the excess is negative; and if $K = 0$ everywhere, the excess is zero. That is, $\sum_{i=1}^{3} \theta_i = \pi$!*

An important special case of this is when one has *constant positive or negative curvature*. When this is normalized to be ± 1, we get the following.

Corollary 12.6.5. *When the curvature is constant ± 1, for a geodesic triangle T, Area$(T) =$ the excess.*

We now complete the proof of Theorem 12.5.4.

Proof. Let $C = X(s)$ be a piecewise smooth closed curve in a geodesic neighborhood of our surface. Using Liouville's formula (see Proposition 12.3.6) for the geodesic curvature (where $F = 0$), we see that

$$\int k_g(s) ds = \int_C (d(\theta) + P du + Q dv),$$

for certain functions $P(u, v)$ and $Q(u, v)$, where $\theta(s)$ is the angle between $X'(s)$ and $X_u(s)$. Since $\int_C d(\theta) + \sum_{i=1}^{n} \alpha_i = 2\pi$, it follows from the local Gauss–Bonnet theorem that excess$(C) = - \int_C (P du + Q dv)$. By Green's theorem, where S is the interior of C,

$$\int_C (P du + Q dv) = \iint_S \left(\frac{\partial Q}{\partial u} - \frac{\partial P}{\partial v} \right) du dv.$$

Therefore, since $\mathrm{excess}(C) = \int K\,dA$, we finally get

$$K(u,v) = -\frac{1}{\sqrt{EG}}\left(\frac{\partial Q}{\partial u} - \frac{\partial P}{\partial v}\right).$$

Taking account of what P and Q are in Proposition 12.3.6, we have

$$K(u,v) = -\frac{1}{2\sqrt{EG}}\left(\frac{\partial}{\partial u}\frac{G_u}{\sqrt{EG}} + \frac{\partial}{\partial v}\frac{E_v}{\sqrt{EG}}\right).$$

Imposing the conditions on the first fundamental form of a geodesic neighborhood, this becomes

$$K(\rho,\theta) = -\sqrt{G}_{\rho,\rho}/\sqrt{G}. \qquad \square$$

This concludes our study of the local Gauss–Bonnet theorem. We now turn to the global Gauss–Bonnet theorem.

Theorem 12.6.6. *Let S be a compact, connected, orientable surface of genus g. Then*

$$\iint K\,dA = 2\pi\chi(S) = 2\pi(2 - 2g).$$

This is rather remarkable. It says after multiplying by a known constant, a differential geometric invariant, namely, total curvature, is equal to a topological invariant, the Euler characteristic.

Proof. Rather than triangulate S here because of the combinatorics, it is more convenient to decompose S into "rectangles" (i.e., quadrilaterals) and in such a way that each rectangle is contained in a geodesic coordinate neighborhood. We then apply the local Gauss–Bonnet Theorem 12.5 to each rectangle. Summing over all the rectangles, for the left-hand term, we get $\iint K\,dA$. All the terms involving k_g drop out because we are integrating over all these edges twice, once in each direction. Letting f, e and v be, respectively, the number of faces, edges and vertices of the decomposition, then we have

$$\iint K\,dA = -2\pi f + \sum_i \theta_i,$$

where the last term is the sum of all the interior angles of all the rectangles involved. But the sum of the interior angles at each vertex is just 2π, so $\sum_i \theta_i = 2\pi v$. Since we have rectangles, each face has four edges. But each edge belongs to two faces. Thus, $4f = 2e$. Since $e = 2f$, $-f = f - e$ and so

$$\iint K \, dA = -2\pi f + 2\pi v = 2\pi(v - e + f). \qquad \square$$

We conclude with some consequences of the global theorem.

Corollary 12.6.7. *Suppose S is a compact connected, orientable surface of curvature $K \geq 0$ and K is positive at some point, then $g = 0$ and S is a diffeomorphic to a sphere. If $K > 0$ and constant, the radius is $\frac{1}{\sqrt{K}}$.*

Proof. Since $K \geq 0$, the global Gauss–Bonnet theorem tells us that $\iint K \, dA = 2\pi(2 - 2g)$. Hence, $2 - 2g > 0$ and $g < 1$. Since g is an integer, it equals zero. If K is constant, then $K \cdot \text{Area}(S) = 4\pi$. Since the area of a sphere of radius r is $4\pi r^2$, the conclusion follows. $\qquad \square$

Along these same lines, we observe that removing a single point from a global surface changes everything. For by stereographic projection, removing a point from the sphere gives a surface with curvature K, identically zero. But S^2 itself cannot have curvature $K \leq 0$. For if it did, then $\iint K \, dA \leq 0$. But this is 4π, a contradiction.

We conclude with some useful facts in the form of exercises all of which except the last depends on the global Gauss–Bonnet theorem.

Exercise 12.6.8. Let S be a compact, connected, smooth, orientable surface of genus g and K be the Gaussian curvature.

1. If $g = 0$, then there must be a point $p \in S$, where $K(p) > 0$.
2. If $g \geq 2$, then there must be a point $p \in S$, where $K(p) < 0$.
3. If $g = 1$, then there are points $p, q \in S$, where $K(p) \geq 0$ and $K(q) \leq 0$. For if not, then all points have positive (respectively negative) curvature, and therefore, the total curvature is positive (respectively negative), a contradiction. Note here that we have quite a general result which does not need any specific computation such as in the exercise on the Gaussian curvature of the torus in the following.
4. Here is a more detailed version of the previous problem. Let S be a Riemannian manifold diffeomorphic with T^2. Show that if

the curvature K is not identically zero, then $S^{\pm}$, the places where $K > 0$, respectively $K < 0$, each have positive area. (In particular, since area is a regular measure, each contains a non-void open set.) Moreover, $X \setminus (X^+ \cup X^-)$ has area zero.

5. Let S be a surface, p_0 be a point where $K(p_0) < 0$ and $D(r, p_0)$ be the geodesic disk centered at p_0 of radius $r > 0$. Show that for r sufficiently small $D(r, p_0)$ has area greater than πr^2 and $D(r, p_0)$ has circumference greater than $2\pi r$.

6. If a compact surface in $\mathbb{R}^3$ has constant curvature K, then K must be positive. By compactness and continuity, there is a point p_0 farthest from the origin. Then $K(p_0)$ cannot be negative or zero.

Appendix A

The Gaussian Distribution

In this appendix, we not only show the very important integral $\int_{-\infty}^{\infty} e^{-x^2} dx$ is finite, but more practically we calculate its value which is of great importance in probability and statistics. We do this by considering an improper double integral and use polar coordinates, the change of variables formula and Fubini's theorem.

Theorem A.0.1 (Euler–Poisson integral).

$$\int_{-\infty}^{\infty} e^{-x^2} dx = \sqrt{\pi}. \tag{A.1}$$

Rescaling the weight function by the factor $\frac{1}{\sqrt{\pi}}$, so that the total area under its graph is 1, gives the *normal distribution* in probability and statistics.

Proof. Changing to polar coordinates and applying the change of variable formula for multiple integrals,

$$\int\int_{\mathbb{R}^2} e^{-(x^2+y^2)} dx dy = \int_0^{\infty} \int_0^{2\pi} e^{-r^2} r dr d(\theta).$$

The right side of this equation is $\int_0^{\infty} e^{-r^2} \int_0^{2\pi} d(\theta) dr$, which is $\pi \int_0^{\infty} e^{-r^2} 2r dr = \pi(1 - e^{-\infty}) = \pi$. On the other hand, $e^{-(x^2+y^2)} = e^{-x^2} e^{-y^2}$. Hence,

$$\int\int_{\mathbb{R}^2} e^{-(x^2+y^2)} dx dy = \int_{\mathbb{R}} e^{-x^2} dx \int_{\mathbb{R}} e^{-y^2} dy = \left[\int_{\mathbb{R}} e^{-x^2} dx \right]^2$$

Again, using the Taylor expansion, we get, for small r,

$$A = \pi r^2 - \frac{\pi}{12} r^4 K(p_0). \qquad (12.4)$$

This gives another intrinsic way of understanding the effect of curvature at a point in terms of the area of small geodesic circles, the area in the tangent space, T_{p_0}, being πr^2. So, here, again we get a correction. If $K(p_0) > 0$, it makes A smaller than the Euclidean area. If $K(p_0) < 0$, it makes A larger.

Just as above, solving for $K(p_0)$ yields

$$K(p_0) = \lim_{r \to 0} \frac{12}{\pi r^4} (\pi r^2 - A),$$

giving a way of calculating the curvature from the areas of small geodesic circles.

Combining equations (12.3) and (12.4) gives us the isoperimetric inequality *with the error term* $\pi^2 r^4 K(p_0)$ (here, we can absorb the r^6 term into ϵ). Compare [10] for the isoperimetric inequality without error term.

Corollary 12.5.6. $4\pi A - l^2 = \pi^2 r^4 K(p_0) + \epsilon$, *where* $\frac{\epsilon}{r^4} \to 0$ *as* $r \to 0$.

12.6   The Gauss–Bonnet Theorem

Here, we come to the most important theorem on the geometry of surfaces. It links geometric and topological invariants and has a number of profound consequences connected with the fundamental concept of curvature. However, to give a complete proof in the general case, we shall have to make some assumptions about the readers' knowledge of the topology of surfaces.

Here are the facts concerning the topology of compact orientable surfaces that we will need.

Let S be a compact connected orientable surface. Then S is homeomorphic to the 2 sphere with a finite number $g \geq 0$ of handles added on. The integer g is called the genus of S. So, for example, if $g = 0$, this is the sphere itself, and for $g = 1$, it is the surface of a donut (both being surfaces of revolution we have already discussed. The former has constant positive curvature and the latter has points by Fubini's theorem. Taking the positive square root of both sides yields the desired integral and thus the conclusion.

In particular,

$$\int \cdots \int_{\mathbb{R}^n} e^{-(x_1^2 + \cdots + x_n^2)} dx_1 \cdots dx_n = \int_{\mathbb{R}^n} e^{-||x||^2} dx = (\pi)^{\frac{n}{2}}.$$

$\square$

We remark that the Euler–Poisson integral is inaccessible by one variable calculus (the antiderivative of e^{-x^2} is not an elementary function). The function e^{-x^2} is known as the *Gaussian* and comes up in many contexts. Its graph is the "bell-shaped curve" of probability and statistics.

Contraction Mappings and Picard's Existence Theorem

Let (X, d) be a metric space and f a self mapping of X. We shall say that f is a *contraction map* if there is a constant $0 < b < 1$ so that for every $x, y \in X$, $d(f(x), f(y)) \le bd(x, y)$.

The point of this definition is the following fixed point theorem of Picard.[1]

Theorem B.0.1. *Let (X, d) be a complete metric space and $f : X \to X$ be a contraction mapping. Then X has a unique point fixed by f.*

One of the earliest and best known fixed point theorems is that of Picard. Either explicitly or implicitly this theorem is the usual way one proves the *local* existence and uniqueness theorem for systems of ordinary differential equations. We remark that it can also be used to prove the inverse function theorem.

Proof.   Choose a point $x_1 \in X$ in an arbitrary manner, and for $n \ge 2$, construct the sequence $x_n \in X$ by $x_{n+1} = f^n(x_1)$. We will first show x_n is a Cauchy sequence.

For $n \ge m$,

$$d(x_n, x_m) = d(f^n(x_1), f^m(x_1)) \le b^m d(f^{n-m}(x_1), x_1).$$

But by the triangle inequality,

$$d(f^{n-m}(x_1), x_1) \le d(f^{n-m}(x_1), f^{n-m-1}(x_1)) + \cdots + d(f(x_1), x_1).$$

---

[1] Emile Picard (1856–1914) was a Professor at Paris and made important contributions to analysis and topology. His book, *Traite d' Analyse*, is well known.

The latter term is less than or equal to $(b^{n-m-1} + \cdots + b + 1)d(f(x_1), x_1)$ which is itself less than or equal to $\sum_{n=0}^{\infty} b^n d(f(x_1), x_1)$. Since $0 < b < 1$, this geometric series converges to $\frac{1}{1-b}d(f(x_1), x_1)$. Since $b^m \to 0$, we see for n and m sufficiently large, given $\epsilon > 0$,

$$d(x_n, x_m) \le b^m \frac{1}{1-b} d(f(x_1), x_1) < \epsilon.$$

Hence, x_n is a Cauchy sequence. Since X is complete, $x_n \to x$ for some $x \in X$. As f is a contraction map, it is (uniformly) continuous. Hence, $f(x_n) \to f(x)$. But as a subsequence, $f(x_n)$ also must converge to x. By the uniqueness of limits, $x = f(x)$.

Now, suppose there was another fixed point $y \in X$. Then

$$d(f(x), f(y)) = d(x, y) \le bd(x, y),$$

so that if $d(x, y) \ne 0$, we conclude $b \ge 1$, a contradiction. Therefore,

$$d(x, y) = 0 \quad \text{and} \quad x = y. \qquad \square$$

B.1   Local Existence Theorems

We now turn to Picard's ODE existence theorem. That is, we consider the problem of existence of solutions of the *most general* first-order ODE,

$$\frac{dy}{dx} = f(x, y), \tag{B.1}$$

where $f : \Omega \to \mathbb{R}$ is any continuous function defined on some domain (open connected set) $\Omega \subseteq \mathbb{R}^2$. As usual, we view y as the dependent variable and x as the independent variable. That is, $y = y(x)$ is the unknown function we seek. Our main purpose is to prove that a wide class of equations of this form have local solutions and that solutions to such initial value problems are unique. As we noted earlier, for example, the *linear equation* $y' = p(x)y + q(x)$, where $p(x)$ and $q(x)$ are continuous functions on some interval, $(\alpha, \beta) \subseteq \mathbb{R}$ has *global* solutions. However, the reader now knows only in rather special cases is it possible to find explicit analytic expressions for these solutions.[2] Here, we will prove that any initial value problem for equation (B.1) always has a unique solution which can be explicitly obtained by an *approximation process*, provided the function f satisfies an additional condition, known as the *Lipschitz condition*. This result due to Emile Picard is of fundamental importance and has applications in a great variety of other mathematical problems. A simple extension to *systems* of first-order ODEs can also be used to establish existence and uniqueness of initial value problems for systems of n ODEs or of an nth-order ODE as above (alternatively, see Chapter 1 of [2]).

Definition B.1.1. Let $(\alpha, \beta) \subseteq \mathbb{R}$ be a given interval and $y(x)$ continuous on (α, β). Given any fixed constant b, an *initial value problem* for our equation is

$$\frac{dy}{dx} = f(x, y), \quad y(a) = b. \tag{B.2}$$

The key idea for solving the initial value problem lies in replacing it by the equivalent *integral equation* in y

$$y(x) = b + \int_a^x f(t, y(t)) dt, \tag{B.3}$$

where $x \in (\alpha, \beta)$ (we call such an equation an integral equation because the unknown function appears in the integrand).

Proposition B.1.2. *A function φ is a solution of the initial value problem on an interval $(\alpha, \beta) \subseteq \mathbb{R}$ if and only if φ is a solution of the integral equation just above on the same interval.*

Proof. If φ is a solution of the initial value problem, then for $t \in (\alpha, \beta)$,

$$\varphi'(t) = f(t, \varphi(t)), \tag{B.4}$$

---

[2]Rudolf Lipschitz (1832–1903) was a Professor at Bonn University who clarified the original existence and uniqueness theorem for ODEs. He also extended Dirichlet's theorem on representing a function of Fourier series, contributed to the calculus of variations and found the number of different ways an integer can be represented as a sum of four squares using integral quaternions.

and $\varphi(a) = b$. Since φ is continuous on (α, β), and f is continuous on Ω, the function $f(t, \varphi(t))$ is continuous on (α, β). Hence, integrating the initial value equation from a to x yields

$$\varphi(x) = b + \int_a^x f(t, \varphi(t))dt,$$

and thus, φ is a solution of the integral equation.

Conversely, if φ satisfies the integral equation, then $\varphi(a) = b$, and differentiating, we find, using the fundamental theorem of calculus, that

$$\varphi'(x) \equiv f(x, \varphi(x)) \text{ on } (\alpha, \beta).$$

Thus, φ is a solution of the initial value problem.   □

Definition B.1.3. A function $f(x, y)$ defined on a set $S \subseteq \mathbb{R}^2$ is said to satisfy a *Lipschitz condition* in y on S if there is a constant $M > 0$ such that

$$|f(x, y_1) - f(x, y_2)| \le M|y_1 - y_2| \tag{B.5}$$

for all (x, y_1), (x, y_2) in S.

Remark B.1.4. A sufficient condition for a function $f(x, y)$ to satisfy a Lipschitz condition on a closed rectangle S in $\mathbb{R}^2$ is the continuity of $\frac{\partial f}{\partial y}(x, y)$ on S, there exist $M > 0$ such that

$$\left|\frac{\partial f}{\partial y}(x, y)\right| \le M \tag{B.6}$$

for all $(x, y) \in S$. At the same time, for (x, y_1), $(x, y_2) \in S$, the mean value theorem implies

$$|f(x, y_1) - f(x, y_2)| = \left|\frac{\partial f}{\partial y}(x, \xi)\right| |y_1 - y_2|,$$

where $y_1 < \xi < y_2$. Now, since $(x, \xi) \in S$, we see that

$$|f(x, y_1) - f(x, y_2)| \le M|y_1 - y_2|.$$

Thus, in the following theorem, it would suffice to assume that f is a C^1 function on Ω.

Theorem B.1.5. *Let Ω be a domain in $\mathbb{R}^2$ and let $f : \Omega \to \mathbb{R}$ be continuous. Let $(a, b) \in \Omega$, and consider the initial value problem*

$$\frac{dy}{dx} = f(x, y), \quad y(a) = b. \tag{B.7}$$

Suppose f satisfies a Lipschitz condition in y on Ω. Then there exists a $\delta > 0$ and a unique solution $\varphi = \varphi(x)$ to the initial value problem for all $|x - a| \leq \delta$.

Proof. Since Ω is open and $v = (a, b) \in \Omega$, there is $r_1 > 0$ such that the open disk $B_{r_1}(v) \subseteq \Omega$. Choose $0 < r < r_1$ so that the closed disk $\overline{B}_r(v) \subseteq B_{r_1}(v)$. Since f is continuous on Ω and so on $\overline{B}_r(v)$, there exists $K > 0$ such that

$$|f(x, y)| \leq K$$

for $(x, y) \in \overline{B}_r(v)$. In addition, since f satisfies a Lipschitz condition in y on Ω and hence on $\overline{B}_r(v)$, there exists $M > 0$ such that

$$|f(x, y_1) - f(x, y_2)| \leq M|y_1 - y_2|$$

for all (x, y_1), (x, y_2) in $\overline{B}_r(v)$. We now choose $0 < \delta < \min\{\frac{r}{K+1}, \frac{1}{M}\}$ such that $\delta < \frac{1}{M}$ and the rectangle $\{(x, y) : |x - a| \leq \delta, \ |y - b| \leq K\delta\} \subseteq \overline{B}_r(v)$.

Let $\mathsf{J} = [a - \delta, a + \delta]$ and consider the metric space $C(\mathsf{J})$ of all continuous real functions $\psi : \mathsf{J} \to \mathbb{R}$ with the metric

$$d(\phi, \psi) = \max_{x \in \mathsf{J}} |\phi(x) - \psi(x)|.$$

This metric is usually called the sup norm metric.

As an exercise, we now ask the reader to verify that for any compact metric space (X, d), the metric space $C(X)$ with the supnorm is a complete metric space. Moreover, any closed subset of a complete metric space is itself complete.

Let

$$\mathcal{C} = \{\psi \in C(\mathsf{J}) : |\psi(x) - b| \leq K\delta\}.$$

The set $\mathcal{C}$ is a non-empty ($\mathcal{C}$ contains the constant function $\psi(x) = b$) closed subset of the complete metric space $(C(\mathsf{J}), d)$. This is itself a complete metric space. In addition, note that if $\psi \in C$, then $|\psi(x) - b| \leq K\delta$ and $(x, \psi(x)) \in \Omega$ for all $x \in J$.

Now, for each $\psi \in C$, consider the mapping F given by

$$F(\psi)(x) = b + \int_a^x f(t, \psi(t))dt,$$

where $x \in J$. First, note that since ψ and f are continuous, all $F(\psi)$ are continuous on J. Moreover, the estimate

$$|F(\psi)(x) - b| = \left| \int_a^x f(t, \psi(t))dt \right| \leq \int_a^x |f(t, \psi(t))| dt$$

$$\leq K \int_a^x dt \leq K\delta$$

shows that $F(\psi) \in C$.

Next, we show that $F : C \to C$ is a contraction mapping of this space. Let $\psi_1, \psi_2 \in C$. Then

$$d(F(\psi_1), F(\psi_2)) = \max_{x \in J} |F(\psi_1)(x) - F(\psi_2)(x)|$$

$$\leq \max_{x \in J} \left| \int_a^x |f(t, \psi_1(t)) - f(t, \psi_2(t))| dt \right|$$

$$\leq \max_{x \in J} \left| \int_a^x M|\psi_1(t) - \psi_2(t)| dt \right|$$

$$\leq \max_{x \in J} \left| M \int_a^x d(\psi_1, \psi_2)dt \right|$$

$$\leq M d(\psi_1, \psi_2) \max_{x \in J} |x - a|$$

$$= M\delta d(\psi_1, \psi_2).$$

Since $M\delta < 1$, F is a contraction mapping. Therefore, from the fixedpoint theorem, it follows that there exists unique $\varphi \in C$ such that $F(\varphi) = \varphi$, i.e.,

$$\varphi(x) = b + \int_a^x f(t, \varphi(t))dt.$$

In other words, φ is a solution of this integral equation. It follows from the proposition that φ is a solution to the initial value problem.

Moreover, φ is the only solution of it. For if ψ were another solution, ψ would also satisfy the integral equation

$$\psi(x) = b + \int_a^x f(t, \psi(t))dt.$$

That is, $F(\psi) = \psi$ which contradicts the uniqueness of φ (i.e., the uniqueness of the fixed point). □

Note that, as a byproduct of the proof, getting our start anywhere, we actually get a sequence of successive approximations to our solution.

More generally, by identical reasoning, we also have the following local existence theorem for systems of n ODEs.

Theorem B.1.6. *Let Ω be a domain in $\mathbb{R}^{n+1}$ and let $\mathbf{f} : \Omega \to \mathbb{R}^n$ be continuous. Let $(a, \mathbf{b}) \in \Omega$, and consider the initial value problem*

$$\frac{d\mathbf{y}}{dx} = \mathbf{f}(\mathbf{x}, \mathbf{y}), \quad \mathbf{y}(a) = \mathbf{b}. \tag{B.8}$$

Suppose $\mathbf{f}$ satisfies a Lipschitz condition in $\mathbf{y}$ on Ω. Then there exists a $\delta > 0$ and a unique solution $\varphi = \varphi(\mathbf{x})$ to the initial value problem for all $|x - a| \leq \delta$.

Finally, as a corollary of this result on systems, we also have a local existence and uniqueness theorem for nth-order equations.

Corollary B.1.7. *Let Ω be a domain in $\mathbb{R}^{n+1}$ and let $f : \Omega \to \mathbb{R}$ be continuous. Let $(a, b_0, \ldots, b_{n-1}) \in \Omega$, and consider the initial value problem*

$$y^n = f(x, y, y', \ldots y^{(n-1)}),$$

where $y(a) = b_0$ $y'(a) = b_1$ and $y^{n-1} = b_{n-1}$. Suppose $f(x, y_0, \ldots, y_{n-1})$ satisfies a Lipschitz condition on Ω. Then there exists a $\delta > 0$ and a unique solution $\varphi = \varphi(x)$ to the initial value problem for all $|x - a| \leq \delta$.

Appendix C

Stokes' Theorem

C.1 Partitions of Unity

Although not necessary, here we will limit ourselves throughout to bounded domains D in $\mathbb{R}^n$.[1]

An important concept in proving Stokes' theorem as well as of importance in its own right is the notion of a *partition of unity*.

Definition C.1.1. The support of a numerical function ϕ defined on D (written $\operatorname{Supp}\phi$) means the closure of $\{x \in D : \phi(x) \neq 0\}$. Let A be a compact subset of $\mathbb{R}^n$ and $\{U_i\}$ be an open covering of A. A family ϕ_i of functions on A is called a *a partition of unity subordinate to the covering* $\{U_i\}$ if $\operatorname{Supp}\phi_i \subseteq U_i$, for all i, if for all $x \in \mathbb{R}^n$, $0 \leq \phi_i(x) \leq 1$, and $\sum_i \phi(x) = 1$ for all $x \in A$.

Note that there could be more than finitely many i, so there would be an issue about taking the above sum. But as we shall see, since A is compact, this problem will not arise. However, in order to deal with the finiteness of the sum in the definition of partition of unity, more generally, instead of assuming that A is compact, one could assume the weaker hypothesis that every covering of A has a locally finite subcover, implying that each point of A has a neighborhood which meets only finitely many elements of the subcover. Such an A is called *paracompact*.

---

[1]Concerning manifolds in general, we remark that there are manifolds where one coordinate neighborhood will not cover it. A good example of this is the surface of a sphere S (in any dimension) where it takes at least two charts to cover S.

A Hausdorff space X is one where for any two distinct points p and q in X, there exist open sets U_p and U_q about p and q, respectively, which are disjoint. This is called a separation axiom and is usually written as T_2. Also, one says X is locally compact if each point $p \in X$ has an open neighborhood with compact closure.

Now, any metric space is Hausdorff and $\mathbb{R}^n$ is locally compact (each point has a compact neighborhood) because closed balls are compact. Since $\mathbb{R}^n$ is a union of the closed balls centered at 0 of arbitrary integer radius and these are compact, $\mathbb{R}^n$ is also σ compact, i.e., a countable union of compact sets. As such, it follows that any covering (of anything) by open sets has a countable subcover.

In order to produce partitions of unity, we shall need the following lemma.

Lemma C.1.2. *Let A and B be non-intersecting sets in $\mathbb{R}^n$ with A compact and B closed. Then there exists a real-valued C^∞ function ϕ on $\mathbb{R}^n$, satisfying the following:*

1. *ϕ is identically 1 on A and identically 0 on B.*
2. *$0 \le \phi(x) \le 1$ everywhere on $\mathbb{R}^n$.*

Proof.   Let a and b be any two real numbers satisfying $0 < a < b$ and $f : \mathbb{R} \to \mathbb{R}$ be defined by $f(x) = \exp(\frac{1}{x-b} - \frac{1}{x-a})$ on (a, b) and zero otherwise, f is well defined on $\mathbb{R}$. Note that $\int_a^b f dx > 0$ since $f > 0$ and continuous on (a, b). Moreover, by L'Hopital's rule, f is C^∞. Let $F(x) = \int_x^b f / \int_a^b f$ for all $x \in \mathbb{R}$. One checks easily that this evidently smooth function F is 0 for $x \ge b$, equals 1 for $x \le a$ and decreases from 1 to 0 on $[a, b]$. Now, let ψ be the radial function defined by $\psi(x_1, \ldots, x_n) = F(||x||)$ for $x \in \mathbb{R}^n$. Then $\psi = 0$ if $||x|| \ge b$, $\psi = 1$ if $||x|| \le a$ and ψ decreases when $a \le ||x|| \le b$. Thus, we now know that for any two concentric balls S and S' with $S' \subset S$ in $\mathbb{R}^n$, there is a C^∞ function ψ on $\mathbb{R}^n$ that is identically 1 on S' and vanishes off S. That is, we have proved the lemma when A and B are closed concentric balls.

Now, consider A and B as in the lemma. Since B is closed, $A \cap B$ is empty and $\mathbb{R}^n$ is Hausdorff, we can cover A by open balls U_i, none of whose closures S_i intersects B. Moreover, within each of these U_i, we can find a smaller concentric sphere S_i' whose union will still cover A. (Prove all this using the metric space properties of $\mathbb{R}^n$.) Since A is compact, a finite number of these will already cover A. Using what we already know about these concentric spheres, for each of these finite number of i, let $\psi_i(x)$ be a smooth functions on $\mathbb{R}^n$ such that $0 \leq \psi_i(x) \leq 1$ for all x, $\psi_i(x) = 1$ on S_i and $\psi_i(x) = 0$ off S_i. Finally, set

$$\phi(x) = 1 - \prod_{i=1}^{n}(1 - \psi_i(x)).$$

Then ϕ is a C^∞ function on $\mathbb{R}^n$ which is 1 on A and vanishes outside of $\cup S_i'$, so certainly vanishes off B. This completes the proof of the lemma. $\square$

Corollary C.1.3. *For any compact subset A of an open set U in $\mathbb{R}^n$ and open covering $\{U_i\}$ of A, there exists a partition of unity subordinate to this covering.*

Proof. Consider an open covering U_i of A. By compactness, A is covered by a finite number $\{U_1, \ldots, U_N\}$ of these. By local compactness of $\mathbb{R}^n$ within each of these open sets U_i, we can find a slightly smaller open V_i so that $\overline{V_i} \subseteq U_i$ which also covers A and is compact, so we can apply the lemma to the pair $(\overline{V_i}, U_i)$ for each i. Since $\overline{V_i}$ is compact and U_i is open, then $\mathbb{R}^n \setminus U_i$ is a closed set disjoint from $\overline{V_i}$. By the lemma, there exist smooth real-valued functions ψ_i on $\cup U_1$ such that $0 \leq \psi_i(x) \leq 1$ everywhere and ψ_i is identically 1 on $\overline{V_i}$ and also ψ_i vanishes off U_i. Let $\psi = \sum_{i=1}^{n} \psi_i$. Then ψ is smooth on $\cup U_1$ and $\psi > 0$ everywhere. Therefore, $\phi_i = \frac{\psi_i}{\psi}$ is a partition of unity. $\square$

We now turn to various related issues.

C.2   Orientability

We say a connected manifold M is *oriented* if for any pair of coordinate neighborhoods U_p of p and U_q of q which have a non-empty intersection, the Jacobean of the transition function, $J = \det(x_p^\partial \alpha / \partial x_q^\beta) > 0$. Since the transition map is a diffeomorphism, $J \neq 0$, then it's either positive or negative. The point is that it doesn't change sign. Thus, orientability is a certain global continuity condition on M.

Intuitively, if M were a surface in 3 space such as a sphere, what this means is that if the normal to the surface at p points outward, then it must also point outward at all points q. Almost all smooth manifolds one encounters are orientable. The classical example of a non-orientable manifold is the Mobius strip.

If the manifold is orientable, we also say that the coordinate systems $x = (x_1, \ldots, x_n)$ about p and $y = (y_1, \ldots, y_n)$ about q define the same orientation if $J > 0$ everywhere and the opposite orientation if $J < 0$ everywhere. Of course, from properties of det, interchanging two coordinate functions will reverse the orientation, so doing this for an even number of times will preserve the orientation. Thus, everything depends on the signature of the permutation of $1, \ldots, n$.

C.3   Exterior Differential Forms

Let M be a submanifold of $\mathbb{R}^n$ of dimension k $(k \leq n)$ and U_i, $i = 1, \ldots, N$ be a finite covering of M by charts $(x_1^i, \ldots, x_k^i)$ on each U_i. Also, let ω^k be a smooth differential k form defined on this chart. Then

$$\omega_i^k = a_{1,\ldots,k}(x)dx_1 \wedge, \ldots \wedge dx_k,$$

where $a_{1,\ldots,k}$ is a smooth real-valued function on U_i and we define the integral of ω_i^k to be

$$\int_{U_i} \omega_i^k = \int_{U_i} a_{1,\ldots,k}(x)dx_1 \ldots dx_k.$$

Note that the convention here is to drop the algebraic $\wedge$ in the integral but keep it when talking purely about differential forms.

To define the integral of ω^k over the whole manifold M, we have to piece these integrals over the various U_i together. To do so, we use a partition of unity ψ_i subordinate to U_i. Then

$$\int_M \omega^k = \int_M \left(\sum_{i=1}^N \right) \psi_i(x)\omega^k(x) = \sum_{i=1}^N \int_{U_i} \psi_i(x)\omega^k(x).$$

This makes sense because each ψ_i vanishes off its U_i. The properties of the partition of unity make the $\int_M \omega^k$ well defined, so that it is independent of the particular finite covering or the particular partition of unity employed.

Before turning to Stokes' theorem, we show that a homeomorphism of Euclidean space taking one open set onto another must take their boundaries onto one another.

Let U be an open set in $\mathbb{R}^n$ and $\varphi : U \to \varphi(U)$ be such a homeomorphism onto another such open set in $\mathbb{R}^n$. Let Ω also be open with $\overline{\Omega} \subset U$. Then

$$\varphi(\partial(\Omega)) = \partial(\varphi(\Omega)).$$

Proof. $\partial(\Omega) = \overline{\Omega} \cap \overline{(U \setminus \Omega)}$. Hence,

$$\varphi(\partial(\Omega)) = \varphi(\overline{\Omega} \cap \overline{(U \setminus \Omega)}) = \varphi(\overline{\Omega}) \cap \varphi(\overline{U \setminus \Omega})$$
$$= \overline{\varphi(\Omega)} \cap \overline{[\varphi(U \setminus \Omega)]} = \overline{\varphi(\Omega)} \cap \overline{[\varphi(U) \setminus \varphi(\Omega)]} = \partial(\varphi(\Omega)).$$
$$\square$$

C.4 Stokes' Theorem

Let D be a compact domain (domain meaning non-trivial interior) in $\mathbb{R}^n$ with boundary, $\partial(D)$ given by an equation $f(x_1, \ldots, x_n) = 0$, where we assume the smooth function f satisfies $\operatorname{grad} f|_{\partial(D)} \neq 0$ at every point. Since D is closed, $\partial(D) \supseteq D$, and since the gradient is non-zero at every point of the boundary, the inverse function theorem tells us $\partial(D)$ is a smooth hypersurface in D. (Hypersurface means its dimension is one less than that of D.)

Now, we put an orientation on $\partial(D)$. For each point p on $\partial(D)$, there is a neighborhood U_p of p in $\mathbb{R}^n$ and we let $\nu(q)$ denote the *outer* normal to this neighborhood of the hypersurface at its various points q. Then these are local coordinates $(n(q), y_2, \ldots, y_n)$ on $D \subseteq \mathbb{R}^n$ so that the coordinates $(y_2, \ldots, y_n)$ are coordinates for $\partial(D)$ and also give an orientation on $\partial(D)$ (induced by the natural orientation on $\mathbb{R}^n$). Suppose ω is a smooth differential $n - 1$ form on D and $i : \partial(D) \to D$ is the injection of $\partial(D)$ into D. Let i^* be the dual map to i on (the finite-dimensional dual) vector space of forms. Then $i^*(\omega)$ is the restriction of ω to $\partial(D)$. An important feature of differential forms is that the operators, differentiation and i^* commute. That is, $d(i^*(\omega)) = i^*(d(\omega))$ for all ω. (d is called the coboundary operator and is very important in algebraic topology. In addition to commuting with i^*, it satisfies $d^2 = 0$.)

Theorem C.4.1.

$$\int_D d(\omega) = \int_{\partial(D)} i^*(\omega).$$

Proof.   Let U_α where $1 \le \alpha \le N$ be a finite covering of D by open balls in $\mathbb{R}^n$ and B^n be the open unit ball in $\mathbb{R}^n$ about 0. Further, let $h_\alpha : B^n \to U_\alpha$ be the natural homeomorphism, namely, translating the center point and stretching or compressing the radius appropriately. The implicit function theorem tells us that by choosing the positive radii of U_α small enough, we can assume that every non-empty intersection, $\partial(D) \cap U_\alpha$ is given by the equation $x_\alpha^1 = 0$, where $(x_\alpha 2, \ldots, x_\alpha^n)$ are local coordinates in U_α. If ϕ_α is a partition of unity subordinate to this covering the the following hold:

1. $\operatorname{Supp} \phi_\alpha \subseteq U_\alpha$ for all x.
2. $0 \le \phi_\alpha(x) \le 1$ for all $x \in \cup_\alpha U_\alpha$.
3. $\sum_{1 \le \alpha \le N} \phi_\alpha(x) = 1$ for all $x \in \cup_\alpha U_\alpha$.

Then since ϕ_α takes numerical values, from equation (3) and the linearity of the integral, it follows that

$$\int_{\partial(D)} i^*(\omega) = \sum_\alpha \int_{\partial(D)} i^*(\phi_\alpha \omega)$$

and

$$\int_D d(\omega) = \sum_\alpha \int_D d(\phi_\alpha \omega).$$

Thus, it suffices to show that for each α,

$$\int_{\partial(D)} i^*(\phi_\alpha \omega) = \int_D d(\phi_\alpha \omega).$$

Since we can reduce the problem to a local one, i.e., to a fixed U_α and ϕ_α, we suppress α as unnecessary baggage! We now write $\phi\omega$ in terms of local coordinates $(x_1, \ldots, x_n)$ on U.

$$\phi\omega = \sum_{k=1}^n (-1)^{k-1} a_k(x) dx_1 \wedge \ldots \hat{dx}_k \wedge dx_n,$$

where $a_k(x)$ are C^∞ functions on D and a symbol with the "hat" above it means that term is to be omitted. Differentiating equation

(2) and using the linearity of differentiation, we get

$$d(\phi\omega) = \left(\sum_{k=1}^{n} \partial a_k(x)/\partial x_k \right) dx_1 \wedge \ldots dx_k \wedge \ldots \wedge dx_n.$$

(the two minuses canceling). We now consider the two possible cases: Either $U \cap \partial(D)$ is empty, or it's not. In the first of these, since $\operatorname{Supp} \phi \subseteq U$, ϕ is identically 0 on $\partial(D)$. Therefore, $\int_{\partial(D)} i^*(\phi\omega) = 0$. Thus, in this case, we must prove $\int_D d(\phi\omega)$ is also 0. However, since $U \cap \partial(D)$ is empty by connectedness, either $U \subseteq D$ or $U \subseteq \mathbb{R}^n \setminus D$. In the latter case, $\int_D d(\phi\omega) = 0$, so we are good here, and we may suppose $U \subseteq D$, and by equation (3), our problem is to show

$$\int_U \left(\sum_{k=1}^{n} \partial a_k(x)/\partial x_k \right) dx_1 \wedge \ldots dx_k \wedge dx_n = 0.$$

But via the smooth coordinate function, h, we have identified U with B^n, so we can regard the domain of our function as all of B^n and consider instead the integral over B^n. Indeed, using the properties of the partition of unity, since ϕ is supported on B^n, we can consider the domain to be all of $\mathbb{R}^{n+1}$ by defining the function to be zero outside B^n (and still maintain smoothness). Then we will show

$$\int_{B^n} \left(\sum_{k=1}^{n} \partial a_k(x)/\partial x_k \right) dx_1 \ldots dx_k \ldots dx_n = 0,$$

which will prove the result in this case.

Let C^n be the unit cube in $\mathbb{R}^n$, namely,

$$C^n = \{(x_1, \ldots, x_n) : |x_k| \leq 1\}.$$

Then $B^n \subseteq C^n$ and we denote by C^{n-1} the one cube in $\mathbb{R}^{n-1}$ which is coordinatized by $(x_2, \ldots, x_n)$.

$$\int_{B^n} \left(\sum_{k=1}^{n} \partial a_k(x)/\partial x_k \right) dx_1 \wedge \ldots dx_k \wedge dx_n$$

$$= \sum_{k=1}^{n} \int_{C^n} \partial a_k(x)/\partial x_k dx_1 \wedge \ldots dx_k \wedge dx_n.$$

But just as before, this last expression is

$$\sum_{k=1}^{n} \int_{C^{n-1}} (-1)^{k-1} \int_{-1}^{1} \partial a_k(x)/\partial x_k dx_1 \wedge \ldots \hat{dx}_k \wedge dx_n.$$

To show it is zero, we evaluate the kth term of this last expression by taking out the constant term (the one with the hat) which equals

$$a(x_1, \ldots, 1, \ldots x_n) - a(x_1, \ldots, -1, \ldots, x_n).$$

Since these are both zero, this proves what we need when the intersection is empty.

We now turn to the case when $U \cap \partial(D)$ is non-empty. Our task is to establish equation (1). Our remarks concerning replacing U by B^n and even C^n as well as the formulas involving C^{n-1} remain valid in the present situation.

Since Stokes' theorem is a generalization of the fundamental theorem of calculus and this hasn't yet made an appearance, we must expect it to in this part of the proof.

As we have established earlier, $B^n \cap \partial(D)$ is given by the equation $x_1 = 0$. This means the integrand over $\int_{B^n \cap \partial(D)}$ lacks dx_1 term, Namely,

$$\phi \omega = (-1)^{n-1} a_1 dx_2 \wedge \ldots \wedge dx_n.$$

So, what we have to prove is

$$\int_{B^n \cap \partial(D)} (-1)^{n-1} a_1 dx_2 \ldots dx_n = \sum_{k=1}^{n} \int_{C^n} \partial(a_k(x))/\partial(x_k) dx_1 \ldots dx_n.$$

For $k \neq 1$, $\partial(a_k(x))/\partial(x_k)$ is a continuous function of x_k. By the fundamental theorem of calculus, we get the following equation:

$$\int_{C^n} \partial(a_k(x))/\partial(x_k) dx_1 \ldots dx_n$$

$$= \int_{C^{n-1}} \left(\int_{-1}^{1} \partial(a_k(x))/\partial(x_k) dx_k \right) dx_1 \ldots \hat{dx}_k dx_n$$

$$= \int_{C^{n-1}} [a_k(1, x_2, \ldots, x_n) - a_k(-1, x_2, \ldots, x_n)] dx_1 \ldots \hat{dx}_k dx_n = 0,$$

since $a_k(1, x_2, \ldots, x_n) = 0 = a_k(-1, x_2, \ldots, x_n)$.

On the other hand, integrating over these various intervals and adding, we get

$$\int_{-1}^{1} \partial(a_1(x))/\partial(x_1)dx_1 = a_1|\partial(D).$$

Substituting this into the equation above tells us

$$\int_{B^n \cap \partial(D)} (-1)^{n-1}a_1 dx_2 \ldots dx_n = \int_{C^{n-1}} (-1)^{n-1}a_1 dx_2 \ldots dx_n,$$

and this completes the proof. □

Real Analytic Functions

We first explain the concept of a series of functions. Let $u_n(x)$, where n is a non-negative integer, be a sequence of real-valued functions, all defined on a common domain $\Omega \subseteq \mathbb{R}$. We shall be interested in the convergence properties of this infinite series (of functions)

$$\sum_{n=0}^{\infty} u_n(x).$$

Definition D.0.1.

1. We shall say a series of numbers $\sum_{n=0}^{\infty} c_n$ converges if the sequence of partial sums $s_k = \sum_{n=0}^{k} c_n$ converges to c. Then we write $\sum_{n=0}^{\infty} c_n = c$. Put another way, convergence means $\lim_{k\to\infty} \sum_{n=k}^{\infty} c_n = 0$. Evidently, if the series converges, c_n tends to 0 as $n \to \infty$.
2. If f is a bounded function on Ω, then we denote by $\|f\|_\Omega$ the $\sup_{x\in\Omega} |f(x)|$.
3. We shall say that this series is *absolutely convergent* at x if $\sum_{n=0}^{\infty} |u_n(x)|$ is convergent.
4. The series is called *uniformly convergent* on Ω if there is a function u on Ω so that $\|\sum_{n=0}^{\infty} u_n - u\|_\Omega \to 0$.

Exercise D.0.2.

1. Show if $\sum_{n=0}^{\infty} c_n$ converges, then c_n tends to 0.
2. Show if $\sum_{n=0}^{\infty} c_n$ converges absolutely, then it converges.
3. Give an example showing the converse of this is false.
4. Show $\|f + g\|_\Omega \leq \|f\|_\Omega + \|g\|_\Omega$.

We first turn to the elegant and effective *Weierstrass comparison theorem.*

Proposition D.0.3. *Let $\sum_{n=0}^{\infty} u_n(x)$ be a series and suppose for some x each $|u_n(x)| \le M_n$ and $\sum_{n=0}^{\infty} M_n$ converges. Then for that x, $\sum_{n=0}^{\infty} u_n(x)$ converges absolutely. If $\sum_{n=0}^{\infty} u_n(x) \le \sum_{n=0}^{\infty} M_n$ for all x in a subinterval I, then $\sum_{n=0}^{\infty} u_n(x)$ converges uniformly and absolutely on I.*

Proof.   Since for each x, $|u_n(x)| \le M_n$ and $\sum_{n=0}^{\infty} M_n$ converges, it follows that $\sum_n^{\infty} |u_k(x)| \le \sum_n^{\infty} M_k$ which tends to zero. Hence, $\sum_{n=0}^{\infty} u_n(x)$ converges absolutely at each point. To see this convergence is uniform at I, just note that we can take the sup over $x \in I$ in the equation above and get $\sum_n^{\infty} |u_k(x)|_I \le \sum_n^{\infty} M_{k_I}$. Thus, $\|\sum_n^{\infty} u_k\|_I$ tends to zero.   □

Corollary D.0.4. *Suppose $\sum_{n=0}^{\infty} a_n(x-a)^n$ is convergent when $|x-a| < r$. Then it is absolutely convergent there as well and is uniformly convergent in any subinterval of the form $[b-a, b+a]$.*

This is because if $|x - a| \le b$, then $|a_n(x-a)^n| \le |a_n b^n|$. Since we are in the interval of convergence, the series $\sum_{n=0}^{\infty} M_n$ converges, where $M_n = |a_n b^n|$, and hence, the conclusion.

We now turn to the study of the most restrictive class of real valued functions of one real-variable, namely, the *real analytic functions.* These are defined as follows.

Definition D.0.5. Let $f : \Omega \to \mathbb{R}$ be a C^∞ function defined on Ω. We shall call it real analytic at $a \in \Omega$ if it is representable as a power series in some sub-neighborhood $(a - r, a + r)$, where $r > 0$. For a fixed f, a_n, called the coefficients, may depend on a as a convergent power series,

$$f(x) = \sum_{n=0}^{\infty} a_n(x - a)^n.$$

If this happens for each $a \in \Omega$, we say that f is analytic on Ω. It might even happen that there is a single power series as above which works for all $a \in \Omega$. For instance, a polynomial $p(x)$ is a trivial example of a convergent power series giving such a real analytic function on all of $\mathbb{R}$. We shall see non-trivial examples of this in the following.

As we shall see shortly, our definition is somewhat redundant since a function defined by a convergent power series must, in fact, be C^∞.

The series representing f is called its Taylor series. When $a = 0$, it is called the Maclaurin series.

We first observe that if $\sum_{n=0}^\infty a_n(x-a)^n$ and $\sum_{n=0}^\infty b_n(x-a)^n$ are two power series with the latter convergent and if for n sufficiently large $|a_n| \le |b_n|$, then the former must also be convergent. This follows from the Weierstrass comparison theorem.

Proposition D.0.6. *If* $f(x) = \sum_{n=0}^\infty a_n(x-a)^n$ *converges in the interval* $(a-r, a+r)$, *then so does* $\sum_{n=1}^\infty na_n(x-a)^{n-1}$.

Proof. Let $x_0 \ne 0$ be a fixed point in the interval where f converges, i.e., $|x_0 - a| < r$. Choose x_1 so that $|x_0 - a| < |x_1 - a| < r$. Then $\frac{|x_0 - a|}{|x_1 - a|} = t$ is positive and less than 1 so that $\lim_{n\to\infty} nt^n = 0$. Then,

$$|na_n(x_0 - a)^{n-1}| = \left| \frac{na_n(x_1 - a)^n}{x_0 - a} \left(\frac{|x_0 - a|^n}{|x_1 - a|^n} \right) \right| = \frac{nt^n}{|x_0 - a|} |a_n x_1^n|.$$

But $\frac{nt^n}{|x_0 - a|} \le 1$ since $\lim_{n\to\infty} nt^n = 0$. This means that

$$|na_n(x_0 - a)^{n-1}| \le |a_n x_1^n|.$$

The Weierstrass comparison theorem now tells us that since the original series converges, $\sum_{n=1}^\infty na_n(x-a)^{n-1}$ is also convergent. □

Theorem D.0.7. *If* $f(x) = \sum_{n=0}^\infty a_n(x-a)^n$ *converges on* $(a-r, a+r)$, *then* f *is differentiable there and* $\frac{df}{dx} = \sum_{n=1}^\infty na_n(x-a)^{n-1}$ *(all on* $(a-r, a+r)$).

Before we prove this theorem, we need one more ingredient. We now prove, under appropriate conditions, that if the derivatives of a sequence of functions converges, then so do the functions.

Proposition D.0.8. *Let* f_n *be a sequence of* C^1 *functions on* $I = (a, b)$. *Suppose* $f_n(x)$ *converges to* $f(x)$ *on all of* I *and* f'_n *converges* *uniformly on I to g. Then the following conditions hold:*

1. *g is continuous on I.*
2. *f_n converges uniformly to a function f.*
3. *$f' = g$, i.e., f is $C^1(I)$.*

Proof. First, g is continuous as the uniform limit of continuous functions. By the fundamental theorem of calculus for each $x \in I$ and each integer n, $f_n(x) - f_n(x_0) = \int_{x_0}^x f_n'(t)dt$. Hence, for any n, m and $x \in I$,

$$|f_n(x) - f_m(x)| \le \int_{x_0}^x |f_n'(t) - f_m'(t)|dt + |f_n(x) - f_m(x_0)|.$$

Let $\epsilon > 0$. Since $f_n(x_0)$ converges and f_n' converges uniformly on I by the Cauchy criterion (the easy way) if n and m are large enough,

$$|f_n(x_0) - f_m(x_0)| < \epsilon$$

and

$$|f_n'(t) - f_m'(t)| < \epsilon$$

Therefore, for all n, m sufficiently large and all $x \in I$,

$$|f_n(x) - f_m(x)| < |x - x_0|\epsilon + \epsilon \le (|b - a| + 1)\epsilon.$$

By completeness, f_n converges uniformly to say f. Again, using the fundamental theorem of calculus for each $x \in I$ and each integer n,

$$f_n(x) - f_n(x_0) = \int_{x_0}^x f_n'(t)dt.$$

Now, let $n \to \infty$. On the left, we get the fundamental theorem of calculus for each $x \in I$ and each integer n, $f(x) - f(x_0)$ and on the right $\int_{x_0}^x g(t)dt$. Therefore,

$$\frac{f(x) - f(x_0)}{x - x_0} = \frac{\int_{x_0}^x g(t)dt}{x - x_0}.$$

Hence, taking limits as $x \to x_0$, we see $f'(x_0) = g(x_0)$ for every $x_0 \in I$. □

Let $\sum_{n=0}^{\infty} u_n(x) = u(x)$ be a series of C^1 functions pointwise convergent to u on I. If $\sum_{n=0}^{\infty} u_n'(x)$ is uniformly convergent, then $\sum_{n=0}^{\infty} u_n'(x)$ converges uniformly to $u'(x)$. That is, the derivative of u is obtained by termwise differentiation of the series $\sum_{n=0}^{\infty} u_n(x)$.

We can now complete the proof of Theorem D.0.7. Using Proposition D.0.6 and Corollary D.0.4, we see that both $\sum_{n=0}^{\infty} a_n(x-a)^n$ and $\sum_{n=1}^{\infty} n a_n(x-a)^{n-1}$ are uniformly convergent on any closed subinterval I of $(a - r, a + r)$. The conclusion now follows from Corollary D.0.7.

Continuing in this way, we see that f is C^∞ and its kth derivative is given by

$$\frac{d^k f}{dx^k}(x) = \sum_{n=k}^{\infty} a_n \frac{n!}{(n-k)!}(x-a)^{n-k}$$

for $x \in (a - r, a + r)$. In particular, each $a_k = \frac{f^k(a)}{k!}$. Thus an analytic function determines and is determined uniquely by its coefficients.

Now, one might ask if it is possible that merely being C^∞ could force a function to be real analytic. The answer is *no* and a good example illustrating this is $f(x) = e^{\frac{-1}{x^2}}$ when $x \neq 0$ and $f(0) = 0$ (see Appendix A). This function is certainly C^∞ on $\mathbb{R} \setminus (0)$. Let us check that it is differentiable at 0. What is $\lim_{x \to 0} \frac{f(x)-f(0)}{x-0}$? This is $\lim_{x \to 0} \frac{1}{x} e^{\frac{-1}{x^2}}$. To evaluate this limit, we let $t = \frac{1}{x}$. Then we get $\lim_{t \to \infty} \frac{t}{e^{t^2}}$ which is zero. Since at points other than zero, we compute $f'(x) = \frac{2}{x^3} e^{\frac{-1}{x^2}}$, we see f is differentiable everywhere and $f'(0) = 0$. Continuing in this way by induction, it follows that for $x \neq 0$, all higher derivatives $f^k(x)$ are of the form $e^{\frac{-1}{x^2}} p(\frac{1}{x})$, where p is some polynomial in $\frac{1}{x}$ and using the same reasoning as in the case of f' and induction, we find that $f^k(0) = 0$ for all $k \geq 1$. Since also $f(0) = 0$, this means that if f were real analytic, its Maclaurin series would be identically zero, which is obviously a contradiction.

We conclude this section with the following necessary and sufficient local condition for real analyticity of a function f.

Theorem D.0.9. *Let f be a C^∞ function on $(a - r, a + r)$, where $r > 0$. Then f is real analytic if and only if there exist positive* constants M and r such that for n sufficiently large

$$\|f\|_{a-r,a+r} \le M(1/r)^n/n!.$$

Perhaps, it is simpler to just say

$$\|f\|_{a-r,a+r} \le Mc^n/n!. \tag{D.1}$$

Proof. If f is C^∞ and satisfies equation (D.1), then by Taylor's theorem with remainder, we see the remainder $R_n = f^n(\bar{x})\frac{(x-a)^n}{n!}$ and $\bar{x} \in (a-r, a+r)$. Taking n large enough to be within the range where the inequality holds, we see $|R_n| \le Mc^n/n!$. Since $e^x = \sum_{n=0}^{\infty} x^n/n!$ converges and therefore the nth term tends to zero, we see R_n tends to zero, so f is analytic.

Conversely, suppose $f(x) = \sum_{n=0}^{\infty} a_n(x-a)^n$ converges for all x in the interval $|x-a| < r$. Then since the nth term tends to zero, there is an $M > 0$ so that $|a_n r^n| \le M$ for all n sufficiently large. As we noted earlier, $f^k(x) = \sum_{n=k}^{\infty} a_n \frac{n!}{(n-k)!}(x-a)^{n-k}$ for $|x-a| < r$. Hence, $|f^k(x)| \le \sum_{n=k}^{\infty} |a_n| r^n \frac{r^k}{r^k} \frac{n!}{(n-k)!}$, i.e.,

$$|f^k(x)| \le \frac{M}{r^k} \sum_{n-k}^{\infty} \frac{n!}{(n-k)!} \left(\frac{|x-a|}{r}\right)^{n-k}.$$

Since taking $x \ne \pm r$, we see $\frac{|x-a|}{r} = t < 1$, we get

$$|f^k(x)| \le \frac{M}{r^k} k!(t^0 + (k+1)t^1 + (k+1)(k+2)/2!t^2 + \cdots).$$

Since $|t| < 1$, this Maclaurin series converges and sums to $\frac{1}{(1-t)^{k+1}}$. One can verify this by calculating the various derivatives of $\frac{1}{(1-t)^{k+1}}$ at 0 and comparing them to the coefficients of the power series

$$t^0 + (k+1)t^1 + (k+1)(k+2)/2!t^2 + \cdots$$

Hence,

$$|f^k(x)| \le \frac{M}{r^k} \frac{k!}{(1-|x-a|/r)^{k+1}},$$

so that taking the sup we get

$$\left\| f^k \right\|_{a-r,a+r} \leq \frac{Mk!}{r^k}.$$

We remark that analytic functions are easier to deal with in the context of complex functions of a complex variable rather than real functions of a real variable (see, e.g., [9] pp. 58–59). In particular, the converse part of Theorem D.0.9 is left as an exercise.

Fourier Series

In this appendix, we make some remarks summarizing how Fourier series[1] fits into orthogonal functions and ultimately differential equations. For details regarding this, see Chapter 10 which discusses the *heat equation* in one space variable.

First, let us consider a locally compact topological space X with a finite positive measure $d\mu$ on it. Of course, X could be a compact space such as a closed interval $[a, b]$. But X could also be the entire real line and $d\mu = e^{-x^2} dx$, where dx is the usual integral on $\mathbb{R}$. This illustrates the interesting situation that finite measure spaces need not be compact! Finiteness of the measure means we can integrate continuous real-valued functions on X and if $f(x) = 1$ for all $x \in X$, then $\int_X f(x)dx = \mu(X)$ a finite positive constant. For convenience, one can divide by this finite positive constant so that in the new measure $\int_X f(x)dx = 1$. This is called *normalization of the measure*.

Let f, g be continuous real-valued functions on X. Then $f + g$ and cf are also continuous. So, this is a real vector space. Observe this vector space in not finite-dimensional unless X is finite. Given a measure, we introduce an inner product on this vector space as follows. For f, g continuous, since fg is also continuous on X and therefore can be integrated, we can take the inner product to be

$$\langle f, g \rangle = \int_X f(x)g(x)d\mu.$$

---

[1] Joseph Fourier (1768–1830) French mathematician and physicist (and also Eygyptologist) used what are now called Fourier series and the Fourier integral to study the mathematical theory of heat conduction via PDEs.

The reader should check that the inner product is a real, symmetric, bilinear form. It is non-degenerate since if $\langle f, g \rangle = 0$ for every continuous g, then $f = 0$. To see this, we just take $g = f$. Then $\int_X f^2(x) d\mu = 0$. But $f^2(x) \geq 0$ and is continuous. We show that $f^2 = 0$. This depends on the regularity of the measure, namely, compact sets have finite measure and non-empty open sets have positive measure.

Lemma E.0.1. *Let g be a continuous function with $g(x) \geq 0$ and $\int_X g(x) d\mu = 0$, then $g = 0$.*

Proof. Suppose $g(x_0) > 0$ at some point $x_0 \in X$. By continuity, there would be a compact neighborhood U_0 of x_0 where it would be everywhere positive. Let m be its greatest lower bound on U_0. Then by compactness of U_0, $m = g(u_0)$ for some $u_0 \in U_0$. Since m and $\mu(U_0)$ are both positive, $\int_X g(x) d\mu \geq m\mu(U_0) > 0$, a contradiction. $\square$

We define $\sqrt{\langle f, f \rangle} = \|f\|$, which we call the L^2 norm of f. Note that if $\|f\| = 0$, then $\langle f, f \rangle = 0$ and by lemma just above $f = 0$. Thus, the only *continuous* function with norm zero is the zero function. Of course, a function which is zero on the irrational numbers and one on the rationals has integral zero, but is itself non-zero.

Since we have this inner product on the space of continuous functions, we get the Cauchy–Schwartz inequality, which states for any f, g,

$$|\langle f, g \rangle|^2 \leq \|f\|^2 \|g\|^2, \tag{E.1}$$

with equality if and only if f, g are linearly dependent.

Proof. Let f, g be continuous functions on X and t be real. Then $\|f - tg\|^2$ is ≥ 0. On the other hand, it is

$$\|f\|^2 - 2\langle f, g \rangle t + \|g\|^2 t^2.$$

Observe that we can assume $g \neq 0$, for if $g = 0$, then the CS inequality just says $0 \leq 0$. Hence, we can assume $\|g\| > 0$ so that we really do have a quadratic polynomial. Since f, g are fixed and t varies and the leading coefficient is positive, this is the graph of a parabola which lies above or perhaps just touches the x-axis. So, it has either just one root, or no roots, but it can't have two roots!

Therefore, the discriminant is ≤ 0 and it has one root if and only if the discriminant is zero. This is if and only if there is a t that makes $\|f - tg\|^2 = 0$, and hence, as we just noted if and only if $f = tg$. When the discriminant is negative, i.e.,

$$4(\langle f, g \rangle)^2 - 4\|f\|^2 \|g\|^2 < 0$$

This gives equation (E.1). $\qquad\qquad\qquad\qquad\qquad\qquad\qquad\square$

Corollary E.0.2.

$$\|(f + g)\| \leq \|f\| + \|g\|.$$

Proof. It is sufficient to show the square of the left side is less than or equal to the square of the right side, i.e., devolves to the CS inequality. $\qquad\qquad\qquad\qquad\qquad\qquad\qquad\qquad\qquad\qquad\square$

This norm gives rise to a distance, namely, $d(f, g) = \|f - g\|$ making the continuous functions on X into a metric space.

Exercise E.0.3. The reader should verify the proof of this corollary and that d is indeed a metric.

Next, we turn to the notion and example of a real Hilbert space which we denote by $L^2(X)$. These are the real-valued square integrable (with respect to $d\mu$) functions on X, i.e., the functions $f : X \to \mathbb{R}$ such that $\int_X f^2(x)dx$ is finite. Of course, if f is continuous, then so is its square, so these functions are in $L^2(X)$. But there are also others which are "limits" of the continuous ones. They are called measurable functions and we need not concern ourselves too much about them, but just note that enlarging our scope to these is similar to completing the rational numbers to get the reals and results in making $L^2(X)$ complete. (This latter fact is called the Riesz–Fischer theorem), and just as the rationals are dense in the reals, the continuous functions on X are dense in $L^2(X)$.

Since the (E.1) inequality holds for the continuous functions and these are dense in $L^2(X)$, it also holds for $L^2(X)$. Thus, for $f, g \in L^2(X)$

$$\left| \int_X fg d\mu \right|^2 \leq \int_X f^2 d\mu \int_X g^2 d\mu. \qquad\qquad (E.2)$$

This means $\int_X fg d\mu$ exists if $f, g \in L^2(X)$.

Now, in $L^2(X)$, we can add functions and multiply a function, a real constant c, and remain in $L^2(X)$. (This involves the linearity properties of the integral.) This means $L^2(X)$ is also a real vector space. For the fact that if $f, g \in L^2(X)$, then $f + g \in L^2(X)$, we argue as follows: $(f + g)^2 = f^2 + 2fg + g^2$ and since $\int_X 2fg d\mu$ if finite, $f + g \in L^2(X)$. Since $L^2(X)$ contains the continuous functions as a subspace, $L^2(X)$ is also *infinite*-dimensional unless X is finite.

Since now we know

$$\left| \int_X f g dx \right|^2 \leq \int_X f^2 dx \int_X g^2 dx, \qquad (E.3)$$

and moreover, this inequality becomes an equality if and only if f and g are linearly dependant, we can now discuss inner products which is what will make $L^2(X)$ a Hilbert space. For $f, g \in L^2(X)$, we define the *inner product* by $\langle f, g \rangle = \int_X f g dx$. Clearly $\langle , \rangle$ is bilinear and symmetric. This makes $L^2(X)$ into a Hilbert space. We need to make a few more observations about it. We can now define a norm

$$\|f\| = \sqrt{\langle f, f \rangle} \geq 0,$$

and this equals 0 if and only if $\int_X f^2 dx = 0$ (of course, if f were continuous, then this would mean $f = 0$). In addition,

$$\|f + g\| \leq \|f\| + \|g\|,$$

and of course,

$$\|cf\| = |c| \, \|f\| \, .$$

By definition, these inequalities make $L^2(X)$ into a normed linear space. As above, given a normed linear space, we get a metric space by taking $d(f, g) = \|f - g\|$. This together with completeness is the definition of a real Hilbert space.

The reader may have noticed that in our account, we didn't state the lemma above for $g \in L^2(X)$. The reason for this is it's not correct! What is correct is that $g = 0$ everywhere except for a subset of X of measure zero! Hence, in order to make d into a metric, we can't just consider functions like f, g but rather *equivalence classes* of these measurable functions which differ on a set of measure zero of X.

However, since here we will restrict ourselves to continuous functions, we need not concern ourselves with this curve ball.

For Fourier series, we consider the sequence of functions on $[0, 2\pi]$,

$$\{1, \cos(x), \sin(x), \cos(2x), \sin(2x) \ldots\},$$

and let f be a continuous (or even piecewise continuous) periodic function of period 2π. Then f can be represented as a Fourier series

$$f(x) = \frac{a_0}{2} + \sum_{n=1}^{\infty} a_n \cos(nx) + \sum_{n=1}^{\infty} b_n \sin(nx),$$

where a_i and b_i, called the Fourier coefficients of f, are to be determined. These coefficients are uniquely determined by f. The real Hilbert space, $L^2[0, 2\pi]$, is tailor-made to help us understand the situation. The first thing to observe is that the sequence of functions on $[0, 2\pi]$ given above is an orthonormal system, i.e.,

$$\langle \phi_n, \phi_m \rangle = \int_a^b \phi_n(x)\phi_m(x)dx/(b-a) = \delta_{n,m},$$

this last term being the Kronecker delta, namely, it's 1 if $n = m$ and 0 otherwise. Due to this, we define the Fourier coefficients of f to be

$$c_n = \int_a^b f(x)\phi_n(x)dx.$$

The question is: How good an approximation is $\sum_{n=0}^{\infty} c_n \phi_n(x)$ and in what sense is it an approximation to f?

Then the following holds for $f \in L^2[0, 2\pi]$:

$$\sum_{n=0}^{\infty} c_n^2 = \int_a^b f(x)^2 dx.$$

This is called the *Parseval equation*. In particular,

$$\sum_{n=0}^{\infty} c_n^2 \leq \int_a^b f(x)^2 dx,$$

which is called Bessel's inequality.

A corollary of Bessel's inequality is the Riemann–Lebesgue Lemma, namely,

$$\lim_{n \to \infty} c_n = 0.$$

Finally, $\lim_{n \to \infty} \sum c_n \phi_n(x) = f(x)$ in $L^2[0, 2\pi]$. This L^2 limit is called the limit in the mean.

For proofs of all these assertions, the reader could consult, for example, [10].

Special Relativity

As we have used Newton's law, $F = ma$, extensively, it would seem to be a good idea to address the question of its applicability. Therefore, we shall make a few remarks concerning special relativity where Newton's law doesn't apply. This is a theory proposed by Albert Einstein in 1905 that describes the propagation of matter (and light) at high speeds which reduces to Newtonian mechanics in the limit as speeds become small. According to special relativity, no wave or particle can travel at a speed greater than the speed, c, of light. Therefore, the usual rules of Newtonian mechanics do not apply when adding velocities that are large. For example, if a particle travels at a speed v with respect to a stationary observer, and another particle travels at a speed v' with respect to the first particle, the speed u of the second particle as seen by the observer is not $v + v'$, as it would be in Newtonian mechanics, but rather

$$\frac{v + v'}{1 + \frac{vv'}{c^2}}.$$

This fact is connected with the relationships between two inertial reference frames involving quantities such, as length and time dilation, and mass increase. These manifest themselves when an observer moving at speed v with respect to a fixed reference point sees length l, time t, distorted from their values l_0 and t_0 at rest. Here, length is *contracted* and time is *dilated* as follows:

$$l = l_0 \sqrt{1 - v^2/c^2},$$
$$t = t_0 / \sqrt{1 - v^2/c^2}.$$

Of course, if v is very small compared to c and so v^2 even smaller when compared to c^2, $1/\sqrt{1 - v^2/c^2}$ is approximately 1. But if v were to approach the same order of magnitude as c, then $1/\sqrt{1 - v^2/c^2}$ would become very significant indeed!

Of particular interest is what happens to mass. Here, m_0 is the rest mass, i.e., the mass when $v = 0$. When the velocity of the observer becomes large, the mass appears to increase, and at the speed of light, it becomes infinite.

$$m = \frac{m_0}{(1 - v^2/c^2)^{1/2}},$$

Thus, in special relativity, time and space are not independent, but are intertwined. We represent the time and space coordinates of a particle in a particular reference frame by a vector in 4 space. That is, we consider Euclidean space–time, $\mathbb{R}^4$, labeling its points (t, x, y, z). The last three being the 3 space coordinates of an event and the first its time coordinate. An "event" being considered as a point in this space. If an observer starting at the origin O moves at constant velocity v (which for simplicity we may take to be along the x-axis) and at time t emits a source of light in all directions with velocity c, then the equation of the spherical wave of light will be $x^2 + y^2 + z^2 - c^2t^2 = 0$, while the location P of the observer will be $(vt, 0, 0)$. What are these coordinates when we take our origin to be P? Einstein's *fundamental postulate of relativity* states that the speed c of light in a vacuum (3×10^{10} km/sec) must be the same in all directions and be *independent* of the speed at which an observer is travelling. This means the equation of a spherical wave of light at P must also be $x'^2 + y'^2 + z'^2 - c^2t'^2 = 0$, where $x' = x - vt$, $y' = y$, $z' = z$ and t' is t plus some incremental quantity to account for the time elapsed. Thus, this postulate requires that linear changes of coordinates must preserve the (*indefinite, but non-degenerate*) quadratic form,

$$x^2 + y^2 + z^2 - c^2t^2.$$

This is called the Lorenz form. Here, because of homogeneity of both space and time, we shall assume we are dealing with $\mathbb{R}$-linear transformations of $\mathbb{R}^4$, which preserve the Lorenz form. These form a non-compact simple Lie group called the Lorenz group. The linear transformation taking such a vector to another reference frame is one that preserves to Lorenz form. If the new inertial reference frame is traveling at speed v relative to the original frame and in the x-direction, then the matrix representation of this (Lorentz) linear transformation is

$$T = \begin{pmatrix} 1/\sqrt{1-v^2/c^2} & -(v/c)/\sqrt{1-v^2/c^2} & 0 & 0 \\ -(v/c)/\sqrt{1-v^2/c^2} & 1/\sqrt{1-v^2/c^2} & 0 & 0 \\ 0 & 0 & 1 & 0 \\ 0 & 0 & 0 & 1 \end{pmatrix}.$$

Note that $\det T = 1$. As one might imagine,

$$1/\sqrt{1-v^2/c^2} \text{ and } -(v/c)/\sqrt{1-v^2/c^2}$$

are actually very familiar functions. They are, respectively, $\cosh(v/c)$ and $\sinh(v/c)$ as can be checked by looking at each of the Taylor series. This makes it clear that in special relativity, it is the ratio, v/c, that counts. It also shows that the 2×2 box in the upper left of the matrix above is a hyperbolic rotation.

The classical Newton's law, $F = ma$, must be modified (since classically it is assumed that m is constant, and as we noted relativistically, this is no longer so). Here, Newton's law takes the form

$$\frac{d}{dt}(mv) = -\operatorname{grad} V,$$

implying that m is no longer constant, but the rate of change of momentum still equals the (variable) force. (This is also true in classical mechanics except that there m is constant!)

As we noted at the beginning of this book, in Newtonian mechanics in any physical process, the total energy, $\frac{1}{2}mv^2 + V$, is preserved, i.e., energy can neither be created nor destroyed. However, here, the situation is completely different — vast amounts of energy can indeed be created by *destroying mass*! This is because the total energy is now $\frac{1}{2}mv^2 + V + m_0c^2$, the extra energy resulting from relativistic effects. It is expressed in the following fateful equation:

$$E = m_0c^2.$$

References

[1]   H. Abbaspour and M. Moskowitz. *Basic Lie Theory*, World Scientific Publishing Co., Singapore, 2007.

[2]   E. Coddington and N. Levinson. *Theory of Ordinary Differential Equations*, McGraw-Hill Book Company, New York, Toronto, London, 1955.

[3]   M. Do Carmo. *Differential Geometry of Curves and Surfaces*, Prentice-Hall, Englewood Cliffs, NJ, 1976.

[4]   I. Farmakis and M. Moskowitz. *A Graduate Course in Algebra, Vols. I and II*, World Scientific Publishing Co., Singapore, 2017.

[5]   I. Gelfand and S. Fomin. *Calculus of Variations*, Prentice-Hall Engelwood Cliffs NJ, 1963.

[6]   F. Gungor. *Lie Symmetry Group Methods for Differential Equations* ArXiv:1901.01543v8 [Math.CA] 1 Jun 2021.

[7]   M. Hirsh, S. Smail, and R. Devaney. *Differential Equations, Dynamical Systems and an Introduction to Chaos*, Third Edition, Academic Press, Oxford UK, 2013.

[8]   W. Klingenberg. *A Course in Differential Geometry*, Springer-Verlag, New York, Heidelberg, Berlin, 1978.

[9]   M. Moskowitz. *A Course in Complex Analysis in One Variable*, World Scientific Publishing Co., Singapore, 2002.

[10] M. Moskowitz and F. Paliogianis. *Functions of Several Real Variables*, World Scientific Publishing Co., Singapore, 2011.

[11] M. Moskowitz and R. Sacksteder. The exponential map and differential equations on real Lie groups, *Journal of Lie Theory*, **13**, 291–306, 2003.

[12]   L. Pontrjagin. *Ordinary Differential Equations*, Addison Wesley Publishing Co., Reading MA, Palo Alto, London, 1962.

[13]   G. Simmons. *Differential Equations with Applications and Historical Notes*, McGraw-Hill Book Company, New York, Toronto, London, 1972.

Index